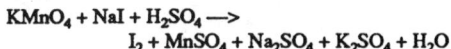

Solutions Manual
for Krimsley's

INTRODUCTORY
CHEMISTRY
2ND EDITION

Victor S. Krimsley
Hartnell College

Brooks/Cole Publishing Company

I(T)P An International Thomson Publishing Company

Pacific Grove · Albany · Bonn · Boston · Cincinnati · Detroit · London · Madrid · Melbourne
Mexico City · New York · Paris · San Francisco · Singapore · Tokyo · Toronto · Washington

Sponsoring Editor: Faith B. Stoddard
Marketing Representative: Jean S. Thompson
Editorial Assistant: Beth Wilbur
Production Coordinator: Tessa A. McGlasson
Cover Design: Vernon T. Boes
Cover Photo: Clark Dunbar, Uniphoto

I(T)P The ITP logo is a trademark under license.

For more information, contact:

Brooks/Cole Publishing Company
511 Forest Lodge Road
Pacific Grove, CA 93950
USA

International Thomson Editores
Campos Eliseos 385, Piso 7
Col. Polanco
11560 México D. F. México

International Thomson Publishing—Europe
Berkshire House 168-173
High Holborn
London WC1V 7AA
England

International Thomson Publishing Gmbh
Königwinterer Strasse 418
53227 Bonn
Germany

Thomas Nelson Australia
102 Dodds Street
South Melbourne, 3205
Victoria, Australia

International Thomson Publishing—Asia
221 Henderson Road #05-10
Henderson Building
Singapore 0315

Nelson Canada
1120 Birchmount Road
Scarborough, Ontario
Canada M1K 5G4

International Thomson Publishing—Japan
Hirakawacho-cho Kyowa Building, 3F
2-2-1 Hirakawacho-cho
Chiyoda-ku, Tokyo 102
Japan

Printed in the United States of America.

5   4   3   2   1

ISBN 0-534-19635-7

Introduction to the
Solutions Manual

This *Solutions Manual* provides all of the worked-out solutions to Krimsley's *Introductory Chemistry,* 2nd edition. While the textbook does provide solutions for the in-chapter problems, this supplement also includes the solutions to the end-of-chapter *Additional Problems,* with the exception of the *Writing Exercises.* The range of acceptable answers to the *Writing Exercises* is too variable to to permit the inclusion of solutions to those exercises.

The purpose of this manual is to allow students to check their answers to the problems in the text, and to see how to solve problems that they are having difficulty with. As stated in the text, success in chemistry depends upon practice in problem-solving. Thus, this manual should only be consulted *after* attempting a given problem. To look up the solution to a problem before trying to solve the problem defeats the purpose of doing problems.

The solutions to the problems in this manual have been worked out by the author of the textbook, using the same approach presented in the text. Thus, there is a consistency between the presentation in the text, and the methods illustrated in this manual.

The author wishes to acknowledge the following individuals for their assistance in the production of this manual: Gloria Langer, who checked the solutions worked out by the author; David Krimsley, who provided technical assistance and prepared the final camera-ready copy; and Faith Stoddard, Lisa Moller, and Beth Wilbur of Brooks/Cole for their encouragement and support.

Contents

CHAPTER 1

Introduction

1.1   The following general problems and their solutions have pervaded the news for over 20 years: (a) the problem of worldwide food shortage can be eased by means of improved agricultural techniques that use chemicals to promote faster growth of crops and by means of gene-splicing techniques that produce heartier crops. (b) The energy shortage can be eased by means of the production of new fuels such as alcohol or shale oil or by means of the development of fusion energy. (c) Air and water pollution can be eased through better control of the release of contaminants and through an increased knowledge of how to remove such contaminants from the environment. In the recent past, the following major news stories appeared: the Chernobyl disaster, the discovery of toxic radon gas levels in certain locations, the development of high-temperature superconductors, the Exxon Valdez oil spill, the depletion of the ozone layer, and the report of a possible cold-fusion reaction.

1.2   (a) 2   (b) 5   (c) 1   (d) 3   (e) 4

1.3   The progress of science depends upon experimentation that is based on earlier findings. By reporting experimental findings in the literature, investigators make their results available to the entire scientific community for further research. The effective use of penicillin as an antibiotic required its synthesis in the laboratory. Although Fleming never synthesized penicillin himself, by reporting his discoveries in the journals, he enabled Florey and Chain to continue his work.

1.4   *Chemistry* is the study of matter and the changes it undergoes.

1.5   Academic chemists generally teach and often engage in research. Industrial chemists are generally involved in the application of chemistry to the production of goods. This may include research and development (R&D) or quality control.

1.6   *Analytical chemistry* concerns itself with either finding out what is present in a given sample, or determining how much of something is present. *Biochemistry* is a study of the chemistry of biological processes. *Inorganic chemistry* concerns the study of substances generally derived from mineral sources. *Organic chemistry* is a study of compounds composed primarily of carbon and hydrogen. *Physical chemistry* concerns the physical properties of matter and the physics of chemical processes.

1.7   An *observation* is a witnessed event. A *hypothesis* is an educated guess used to explain an observation. *Controlled experiments*, designed to alter only one variable at a time, are used to gather information and to test hypotheses. A *law* is a statement of a regularity in observations. A *theory* is a hypothesis that is supported by a large body of evidence.

1.8   A *model* is a concrete visual image, used to describe observations.

1.9   When a considerable body of evidence in support of a hypothesis has been gathered, the hypothesis may become a theory.

1

1.10 Ten plastic products include: plastic bottles, plastic office organizers, casings for electronic equipment such as stereos and telephones, CDs, computer disc holders, plastic food containers, plastic bags, electrical switchplates, credit cards, audio and vidio cassette tapes, toothbrushes and hairbrushes, eyeglass lenses and frames, laundry baskets, clothing hangers, plastic children's furniture, toys, picture frames, vinyl coverings, shower curtains, waste baskets and garbage cans–the list goes on.

1.11 As clothing ages, the dyes in the fabrics begin to fade. Thus, the age of clothing can be ascertained by the intensity of the colors–more faded colors correspond to older clothing. Some other observations that can be used to date clothing would include yellowing (older fabrics tend to yellow), threadbare areas (as are found on collars, elbows, and knees), and holes.

CHAPTER 2

The Classification of Matter

2.1 A solid has a definite shape and volume and tends to resist any change in its shape. A liquid flows freely, having a definite volume, but its shape changes to fit the shape of its container. A gas has no definite shape or volume but expands or contracts to fill exactly the container it is in. Thus, gases can be compressed easily. The solid and liquid states are known as the "condensed" phases.

2.2 (a) homogeneous (b) heterogeneous (c) homogeneous (d) heterogeneous (e) heterogeneous (f) heterogeneous (g) heterogeneous (h) homogeneous (i) heterogeneous

2.3 Homogeneous matter is uniform throughout. Heterogeneous matter is not uniform.

2.4 A pure substance has a fixed and definite composition. A homogeneous mixture may be prepared in different proportions.

2.5 A compound can be decomposed into simpler substances. An element cannot be decomposed into simpler substances by ordinary means.

2.6 A: compound; B: insufficient information; C: compound; D: element; E: insufficient information

2.7 (a) A molecule is the smallest unit of a pure substance that possesses all of the characteristic properties of the substance. An atom is a building block from which molecules are made.
(b) An element is composed of only one kind of atom. A compound is composed of two or more kinds of atoms.
(c) Diatomic substances have two atoms per molecule. Monatomic substances have one atom per molecule. Oxygen and hydrogen are diatomic elements. Helium and argon are monatomic elements. Carbon monoxide is a diatomic compound. It is not possible to have a monatomic compound, since a compound must contain atoms of at least two different elements.

2.8 (a) HCl (b) N_2O_5 (c) C_6H_6 (d) H_2SO_4 (e) $NaCl$ (f) Ne (g) $MgBr_2$ (h) Al_2S_3

2.9 (a) 4 (b) 12 (c) 2 (d) 9

2.10 When a substance undergoes a phase change, such as melting or boiling, the fixed composition of the substance does not change.

2.11 A chemical change involves a change in the fixed composition of the reactants, as new substances with different composition are formed. A physical change does not involve any change in the composition of the substance.

3

2.12 (a) physical property (b) physical property (c) physical property (d) chemical property (e) physical property (f) chemical property (g) physical property

2.13 A *solid* has a fixed volume and is resistant to changes in shape; examples include an ice cube and a gold bar. A *liquid* has a fixed volume but flows freely; examples include ethyl alcohol and carbon tetrachloride. A *gas* expands and contracts to fit exactly in its container; examples include air and helium.

2.14 (a) solid (b) liquid (c) solid (d) gas (e) liquid (f) gas

2.15 (b) homogeneous (b) heterogeneous (c) homogeneous (d) heterogeneous

2.16 (a) heterogeneous (b) homogeneous (c) heterogeneous (d) heterogeneous (e) homogeneous (f) heterogeneous (g) heterogeneous

2.17 Pure substances: a, b, d, e, h, i; Homogeneous mixtures: c, f, g, j

2.18 An *element* is a pure substance that cannot be broken down into simpler substances by ordinary means. A *compound* is a pure substance that can be decomposed into simpler substances.

2.19 (a) He (b) Si (c) Ag (d) Ar (e) U (f) S (g) Cu (h) K (i) Au (j) Fe

2.20 (a) nitrogen (b) carbon (c) sodium (d) chlorine (e) sulfur (f) hydrogen (g) oxygen (h) magnesium (i) calcium (j) lead

2.21 Elements: b, c, e, g, h; Compounds: a, d, f

2.22 Refer to Figure 2-4

2.23 Refer to Figure 2-5. (a) 1 He atom (b) 1 C atom, 4 H atoms (c) 1 C atom, 2 O atoms (d) 2 O atoms

2.24 (a) 3 hydrogen atoms, 1 phosphorus atom, and 4 oxygen atoms
(b) 12 carbon atoms, 22 hydrogen atoms, and 11 oxygen atoms
(c) 1 hydrogen atom, 1 nitrogen atom, and 3 oxygen atoms
(d) 4 phosphorus atoms and 10 oxygen atoms
(e) 3 carbon atoms, 2 hydrogen atoms, and 6 fluorine atoms

2.25 (a) Li_2CO_3 (b) K_2S (c) Al_2O_3 (d) $C_{12}H_{22}O_{11}$ (e) CH_3NO_2

2.26 (a) 3 (b) 6 (c) 9 (d) 1 nitrogen atom and 3 oxygen atoms

2.27 A *physical change* is one in which there is no change in the fixed composition of the substance undergoing the change. For example, when ice melts, water changes from its solid state to its liquid state.

2.28 A *chemical change* is one that involves the production of new substances. When gasoline burns with oxygen, the compounds carbon dioxide and water are formed.

2.29 *The Law of Conservation of Mass* says that matter can neither be created nor destroyed. This means that all of the matter present in the reactants of a chemical change must be found somewhere in the products, and all of the matter found in the products must have come from reactants.

2.30 A burning log gives off energy. Photosynthesis requires absorption of energy.

2.31 A *physical property* is one whose observation does not involve a change in the fixed composition of the substance; examples include melting point, color, and physical state.

2.32 A *chemical property* is one whose observation requires that the substance undergoes a chemical change; examples include flammability and the ability to be oxidized.

2.33 Physical properties: a, b, c, d, g; Chemical properties: e, f

2.34 Carbon dioxide

2.35 (a) The *disappearance* of mothballs is not necessarily a violation of the Law of Conservation of Mass. Just because we cannot *see* the matter does not mean it does not exist.
   (b) The mothballs might have formed a vapor, similar to the water vapor formed from evaporation.
   (c) When one opens a drawer or closet containing mothballs, the odor of the mothballs can be observed. In order to smell the odor of the mothballs, molecules of the substance present in mothballs must be in the air (or the mothballs must have vaporized).
   (d) The disappearance is a physical change.

2.36 $Al_2(SO_4)_3$. The metals combine with half as many sulfates as nitrates. For aluminum, half of 1:3 would be 1:1.5. This corresponds to a 2:3 ratio.

2.37 The mass of the contents at the end of the reaction must equal the mass at the beginning of the reaction. If this were not true, the Law of Conservation of Mass would be violated.

2.38 When the clay is heated and then cooled to the original temperature, the properties of the clay have been permanently altered. Thus, a chemical change must occur. If a physical change were involved, the properties of the clay would return to their original state.

2.39 When a candle burns, oxygen is consumed. Thus, lighting a candle will deplete the oxygen supply.

CHAPTER 3

Measurement

3.1   (a) Exact (b) Measured (c) Exact (d) Measured

3.2   (a) 3   (b) 2   (c) 2 (The last zero holds the decimal.) (d) 4   (e) 3   (f) 3
   (g) 2 (The leading zeros hold the decimal.)
   (h) 3 (The leading zeros hold the decimal, but the last zero must be measured and is therefore significant.)
   (i) 4   (j) 3   (k) 8 (Since the last zero is significant, so are all those in between.)
   (l) 4   (m) 2   (n) 3

3.3   (a) 17.6 (b) 0.00627 (c) 36.0 (d) 36 (e) 40 (4×10^1) (f) 7.0×10^{-6}

3.4   4 significant figures; 6.100×10^4

3.5   (a) 43.6 (remains as is)   (b) 45 (rounded off from 45.28)
   (c) 12.7 (rounded off from 12.70)   (d) 0.8 (rounded off from 0.843)
   (e) 1840 (rounded off from 1838.113)

3.6   (a) 19.1 (b) 3.88 (c) 2.248 (d) 1100 (e) 2.0 (f) 5

3.7   (a) $(11.62)(0.55) = 6.4$ (b) $\dfrac{57.80}{99.0} = 0.584$

3.8   (a) $3.0 \text{ ft}\left(\dfrac{12 \text{ in.}}{1 \text{ ft}}\right) = 36 \text{ in.}$ (b) $54 \text{ in.}\left(\dfrac{1 \text{ ft}}{12 \text{ in.}}\right) = 4.5 \text{ ft}$ (c) $2.0 \text{ yd}\left(\dfrac{3 \text{ ft}}{1 \text{ yd}}\right)\left(\dfrac{12 \text{ in.}}{1 \text{ ft}}\right) = 72 \text{ in.}$

3.9   (a) 1 m = 100 cm; 1 L = 1000 mL

3.10 (a) 1 kg = 1000 g   (b) 1 mm = 10^{-3} m or 1000 mm = 1 m
   (c) 1 cL = 10^{-2} L or 100 cL = 1 L   (d) 1 kJ = 1000 J
   (e) 1 μsec = 10^{-6} sec or 10^6 μsec = 1 sec   (f) 1 mg = 10^{-3} g or 1000 mg = 1 g
   (g) 1 MW = 10^6 W   (h) 1 μL = 10^{-6} L or 10^6 μL = 1 L

3.11 (a) Exact (b) Measured (c) Exact (d) Exact (e) Measured

3.12 (a) $42.6 \text{ cm}\left(\dfrac{1 \text{ m}}{100 \text{ cm}}\right) = 0.426 \text{ m}$ (b) $1.5 \text{ km}\left(\dfrac{1000 \text{ m}}{1 \text{ km}}\right) = 1500 \text{ m } (1.5 \times 10^3 \text{ m})$

   (c) $502 \text{ m}\left(\dfrac{1 \text{ km}}{1000 \text{ m}}\right) = 0.502 \text{ km}$ (d) $0.372 \text{ m}\left(\dfrac{100 \text{ cm}}{1 \text{ m}}\right) = 37.2 \text{ cm}$

3.12 (continued)

(e) $943 \text{ mm} \left(\dfrac{1 \text{ m}}{1000 \text{ mm}} \right) = 0.943 \text{ m}$ (f) $10.0 \text{ cm} \left(\dfrac{1 \text{ in.}}{2.54 \text{ cm}} \right) = 3.94 \text{ in.}$

(g) $3.86 \text{ in.} \left(\dfrac{2.54 \text{ cm}}{1 \text{ in.}} \right) = 9.80 \text{ cm}$ (h) $15.0 \text{ m} \left(\dfrac{100 \text{ cm}}{1 \text{ m}} \right) = 1500 \text{ cm} \ (1.50 \times 10^3 \text{ cm})$

3.13 (a) km $\longrightarrow$ m $\longrightarrow$ mm

$3.23 \text{ km} \left(\dfrac{1000 \text{ m}}{1 \text{ km}} \right) \left(\dfrac{1000 \text{ mm}}{1 \text{ m}} \right) = 3{,}230{,}000 \text{ mm} \ (3.23 \times 10^6 \text{ mm})$

(b) km $\longrightarrow$ m $\longrightarrow$ cm

$1.5 \text{ km} \left(\dfrac{1000 \text{ m}}{1 \text{ km}} \right) \left(\dfrac{100 \text{ cm}}{1 \text{ m}} \right) = 150{,}000 \text{ cm} \ (1.5 \times 10^5 \text{ cm})$

(c) ft $\longrightarrow$ in. $\longrightarrow$ cm

$1.40 \text{ ft} \left(\dfrac{12 \text{ in.}}{1 \text{ ft}} \right) \left(\dfrac{2.54 \text{ cm}}{1 \text{ in.}} \right) = 42.7 \text{ cm}$

(d) cm $\longrightarrow$ in. $\longrightarrow$ ft

$125 \text{ cm} \left(\dfrac{1 \text{ in.}}{2.54 \text{ cm}} \right) \left(\dfrac{1 \text{ ft}}{12 \text{ in.}} \right) = 4.10 \text{ ft}$

(e) yd $\longrightarrow$ ft $\longrightarrow$ in. $\longrightarrow$ cm

$3.20 \text{ yd} \left(\dfrac{3 \text{ ft}}{1 \text{ yd}} \right) \left(\dfrac{12 \text{ in.}}{1 \text{ ft}} \right) \left(\dfrac{2.54 \text{ cm}}{1 \text{ in.}} \right) = 293 \text{ cm}$

(f) ft $\longrightarrow$ in. $\longrightarrow$ cm $\longrightarrow$ m

$10.0 \text{ ft} \left(\dfrac{12 \text{ in.}}{1 \text{ ft}} \right) \left(\dfrac{2.54 \text{ cm}}{1 \text{ in.}} \right) \left(\dfrac{1 \text{ m}}{100 \text{ cm}} \right) = 3.05 \text{ m}$

3.14 (a) m $\longrightarrow$ cm $\longrightarrow$ in.

$1.00 \text{ m} \left(\dfrac{100 \text{ cm}}{1 \text{ m}} \right) \left(\dfrac{1 \text{ in.}}{2.54 \text{ cm}} \right) = 39.4 \text{ in.}$

(b) km $\longrightarrow$ m $\longrightarrow$ cm $\longrightarrow$ in. $\longrightarrow$ ft $\longrightarrow$ mi

$1.00 \text{ km} \left(\dfrac{1000 \text{ m}}{1 \text{ km}} \right) \left(\dfrac{100 \text{ cm}}{1 \text{ m}} \right) \left(\dfrac{1 \text{ in.}}{2.54 \text{ cm}} \right) \left(\dfrac{1 \text{ ft}}{12 \text{ in.}} \right) \left(\dfrac{1 \text{ mi}}{5280 \text{ ft}} \right) = 0.621 \text{ mi}$

3.15 kg $\longrightarrow$ g $\longrightarrow$ lb

$$1.00 \text{ kg}\left(\frac{1000 \text{ g}}{1 \text{ kg}}\right)\left(\frac{1 \text{ lb}}{454 \text{ g}}\right) = 2.20 \text{ lb}$$

3.16 (a) $1235 \text{ g}\left(\frac{1 \text{ kg}}{1000 \text{ g}}\right) = 1.235 \text{ kg}$ (b) $3.45 \text{ g}\left(\frac{1000 \text{ mg}}{1 \text{ g}}\right) = 3450 \text{ mg } (3.45 \times 10^3 \text{ mg})$

(c) $598 \text{ mg}\left(\frac{1 \text{ g}}{1000 \text{ mg}}\right) = 0.598 \text{ g}$ (d) $3.00 \text{ lb}\left(\frac{454 \text{ g}}{1 \text{ lb}}\right) = 1360 \text{ g } (1.36 \times 10^3 \text{ g})$

(e) $7.00 \text{ kg}\left(\frac{1000 \text{ g}}{1 \text{ kg}}\right)\left(\frac{1 \text{ lb}}{454 \text{ g}}\right) = 15.4 \text{ lb}$ (f) $3.50 \text{ lb}\left(\frac{454 \text{ g}}{1 \text{ lb}}\right)\left(\frac{1 \text{ kg}}{1000 \text{ g}}\right) = 1.59 \text{ kg}$

(g) $0.620 \text{ lb}\left(\frac{454 \text{ g}}{1 \text{ lb}}\right)\left(\frac{1000 \text{ mg}}{1 \text{ g}}\right) = 281,000 \text{ mg } (2.81 \times 10^5 \text{ mg})$

3.17 (a) $525 \text{ mL}\left(\frac{1 \text{ L}}{1000 \text{ mL}}\right) = 0.525 \text{ L}$

(b) $4.95 \text{ L}\left(\frac{1000 \text{ mL}}{1 \text{ L}}\right) = 4950 \text{ mL } (4.95 \times 10^3 \text{ mL})$

(c) $1.32 \text{ qt}\left(\frac{946 \text{ mL}}{1 \text{ qt}}\right) = 1250 \text{ mL } (1.25 \times 10^3 \text{ mL})$

(d) $2.25 \text{ qt}\left(\frac{946 \text{ mL}}{1 \text{ qt}}\right)\left(\frac{1 \text{ L}}{1000 \text{ mL}}\right) = 2.13 \text{ L}$

(e) $10.8 \text{ gal}\left(\frac{4 \text{ qt}}{1 \text{ gal}}\right)\left(\frac{946 \text{ mL}}{1 \text{ qt}}\right)\left(\frac{1 \text{ L}}{1000 \text{ mL}}\right) = 40.9 \text{ L}$

(f) $45.3 \text{ L}\left(\frac{1000 \text{ mL}}{1 \text{ L}}\right)\left(\frac{1 \text{ qt}}{946 \text{ mL}}\right)\left(\frac{1 \text{ gal}}{4 \text{ qt}}\right) = 12.0 \text{ gal}$

3.18 (a) $V = (12.0 \text{ cm})(8.00 \text{ cm})(9.50 \text{ cm}) = 912 \text{ cm}^3$

(b) $V = 912 \text{ cm}^3\left(\frac{1 \text{ mL}}{1 \text{ cm}^3}\right) = 912 \text{ mL}$

(c) $V = 912 \text{ mL}\left(\frac{1 \text{ L}}{1000 \text{ mL}}\right) = 0.912 \text{ L}$

3.19 (a) $\dfrac{4.0 \text{ g}}{2.5 \text{ mL}} = 1.6 \text{ g/mL}$ (b) $\dfrac{34.0 \text{ g}}{2.50 \text{ mL}} = 13.6 \text{ g/mL}$

    (c) $\dfrac{59.2 \text{ g}}{75.0 \text{ mL}} = 0.789 \text{ g/mL}$ (d) $\dfrac{15.6 \text{ g}}{5.00 \text{ mL}} = 3.12 \text{ g/mL}$

    (e) $\dfrac{8.69 \text{ g}}{11.0 \text{ mL}} = 0.790 \text{ g/mL}$

3.20 (a) $11.5 \text{ mL}\left(\dfrac{0.85 \text{ g}}{1 \text{ mL}}\right) = 9.8 \text{ g}$ (b) $12 \text{ g}\left(\dfrac{1 \text{ mL}}{1.59 \text{ g}}\right) = 7.5 \text{ mL}$

    (c) $65.0 \text{ mL}\left(\dfrac{0.879 \text{ g}}{1 \text{ mL}}\right) = 57.1 \text{ g}$ (d) $27.0 \text{ g}\left(\dfrac{1 \text{ mL}}{1.48 \text{ g}}\right) = 18.2 \text{ mL}$

    (e) $1.00 \text{ lb}\left(\dfrac{454 \text{ g}}{1 \text{ lb}}\right)\left(\dfrac{1 \text{ mL}}{13.6 \text{ g}}\right) = 33.4 \text{ mL}$ (f) $245 \text{ mL}\left(\dfrac{0.789 \text{ g}}{1 \text{ mL}}\right) = 193 \text{ g}$

3.21 (a) $\text{sp gr} = \dfrac{\text{Density of bromine}}{\text{Density of water}} = \dfrac{3.12 \text{ g}/\text{mL}}{1.00 \text{ g}/\text{mL}} = 3.12$

    (b) $\text{sp gr} \times \text{Density of water} = \text{Density of bromine}$
       $3.12 \times 62.4 \text{ lb/ft}^3 = 195 \text{ lb/ft}^3$

    (c) $\text{sp gr} = \dfrac{\text{Density of carbon tetrachloride}}{\text{Density of water}} = \dfrac{99.5 \text{ lb}/\text{ft}^3}{62.4 \text{ lb}/\text{ft}^3} = 1.59$

    (d) $\text{sp gr} \times \text{Density of water} = \text{Density of carbon tetrachloride}$
       $1.59 \times 1.00 \text{ g/mL} = 1.59 \text{ g/mL}$

3.22 (a) $T_K = 20 + 273;$ $T_K = 293 \text{ K}$ (b) $T_K = 145 + 273;$ $T_K = 418 \text{ K}$
    (c) $T_K = -78 + 273;$ $T_K = 195 \text{ K}$ (d) $T_K = -223 + 273;$ $T_K = 50 \text{ K}$
    (e) $298 = T_C + 273;$ $T_C = 25°C$ (f) $577 = T_C + 273;$ $T_C = 304°C$
    (g) $100 = T_C + 273;$ $T_C = -173°C$ (h) $23 = T_C + 273;$ $T_C = -250°C$

3.23 (a) $87 = 1.8T_C + 32;$ $T_C = 31°C$ (b) $68 = 1.8T_C + 32;$ $T_C = 20°C$
    (c) $-40 = 1.8T_C + 32;$ $T_C = -40°C$ (d) $0 = 1.8T_C + 32;$ $T_C = -18°C$
    (e) $-48 = 1.8T_C + 32;$ $T_C = -44°C$ (f) $-104 = 1.8T_C + 32;$ $T_C = -76°C$
    (g) $932 = 1.8T_C + 32;$ $T_C = 500°C$ (h) $400 = 1.8T_C + 32;$ $T_C = 204°C$

3.24 (a) $T_K = 31 + 273;$ $T_K = 304 \text{ K}$ (b) $T_K = 20 + 273;$ $T_K = 293 \text{ K}$
    (c) $T_K = -40 + 273;$ $T_K = 233 \text{ K}$ (d) $T_K = -18 + 273;$ $T_K = 255 \text{ K}$
    (e) $T_K = -44 + 273;$ $T_K = 229 \text{ K}$ (f) $T_K = -76 + 273;$ $T_K = 197 \text{ K}$
    (g) $T_K = 500 + 273;$ $T_K = 773 \text{ K}$ (h) $T_K = 204 + 273;$ $T_K = 477 \text{ K}$

3.25 (a) $T_F = 1.8(95) + 32;$ $T_F = 203°F$ (b) $T_F = 1.8(60) + 32;$ $T_F = 140°F$
    (c) $T_F = 1.8(20) + 32;$ $T_F = 68°F$ (d) $T_F = 1.8(-40) + 32;$ $T_F = -40°F$
    (e) $T_F = 1.8(-70) + 32;$ $T_F = -94°F$ (f) $T_F = 1.8(37) + 32;$ $T_F = 99°F$

3.26 (a) $59 = 1.8T_C + 32;$ $T_C = 15°C;$ $T_K = 288$ K
          (b) $-13 = 1.8T_C + 32;$ $T_C = -25°C;$ $T_K = 248$ K
          (c) $257 = 1.8T_C + 32;$ $T_C = 125°C;$ $T_K = 398$ K

3.27 (a) $25.0 \text{ cal} \left(\dfrac{4.184 \text{ J}}{1 \text{ cal}} \right) = 105$ J   (b) $0.575 \text{ cal} \left(\dfrac{4.184 \text{ J}}{1 \text{ cal}} \right) = 2.41$ J

          (c) $1.43 \text{ J} \left(\dfrac{1 \text{ cal}}{4.184 \text{ J}} \right) = 0.342$ cal   (d) $1.75 \text{ kcal} \left(\dfrac{4.184 \text{ kJ}}{1 \text{ kcal}} \right) = 7.32$ kJ

          (e) $325 \text{ cal} \left(\dfrac{4.184 \text{ J}}{1 \text{ cal}} \right) \left(\dfrac{1 \text{ kJ}}{1000 \text{ J}} \right) = 1.36$ kJ

3.28 temperature, heat, heat, temperature

3.29 (a) $4.50 \text{ yd}^2 \left(\dfrac{3 \text{ ft}}{1 \text{ yd}} \right)^2 = 4.5 \text{ yd}^2 \left(\dfrac{3^2 \text{ ft}^2}{1^2 \text{ yd}^2} \right) = 40.5 \text{ ft}^2$

          (b) $3.50 \text{ m}^3 \left(\dfrac{10 \text{ dm}}{1 \text{ m}} \right)^3 = 3.50 \text{ m}^3 \left(\dfrac{10^3 \text{ dm}^3}{1^3 \text{ m}^3} \right) = 3.50 \times 10^3 \text{ dm}^3$

          (c) $3.50 \text{ m}^3 \left(\dfrac{100 \text{ cm}}{1 \text{ m}} \right)^3 \left(\dfrac{1 \text{ mL}}{1 \text{ cm}^3} \right) \left(\dfrac{1 \text{ L}}{1000 \text{ mL}} \right)$

          $= 3.50 \text{ m}^3 \left(\dfrac{100^3 \text{ cm}^3}{1 \text{ m}^3} \right) \left(\dfrac{1 \text{ mL}}{1 \text{ cm}^3} \right) \left(\dfrac{1 \text{ L}}{1000 \text{ mL}} \right) = 3.50 \times 10^3 \text{ L}$

          (d) $7.25 \text{ ft}^3 \left(\dfrac{12 \text{ in.}}{1 \text{ ft}} \right)^3 \left(\dfrac{2.54 \text{ cm}}{1 \text{ in.}} \right)^3 \left(\dfrac{1 \text{ mL}}{1 \text{ cm}^3} \right) \left(\dfrac{1 \text{ L}}{1000 \text{ mL}} \right)$

          $= 7.25 \text{ ft}^3 \left(\dfrac{12^3 \text{ in.}^3}{1^3 \text{ ft}^3} \right) \left(\dfrac{2.54^3 \text{ cm}^3}{1^3 \text{ in.}^3} \right) \left(\dfrac{1 \text{ mL}}{1 \text{ cm}^3} \right) \left(\dfrac{1 \text{ L}}{1000 \text{ mL}} \right) = 205$ L

          $1 \text{ L} = 1 \text{ dm}^3$

3.30 $1.00 \text{ ft}^3 \left(\dfrac{12 \text{ in.}}{1 \text{ ft}} \right)^3 \left(\dfrac{2.54 \text{ cm}}{1 \text{ in.}} \right)^3 \left(\dfrac{1 \text{ mL}}{1 \text{ cm}^3} \right) \left(\dfrac{1 \text{ qt}}{946 \text{ mL}} \right) \left(\dfrac{1 \text{ gal}}{4 \text{ qt}} \right)$

          $= 1.00 \text{ ft}^3 \left(\dfrac{12^3 \text{ in.}^3}{1^3 \text{ ft}^3} \right) \left(\dfrac{2.54^3 \text{ cm}^3}{1^3 \text{ in.}^3} \right) \left(\dfrac{1 \text{ mL}}{1 \text{ cm}^3} \right) \left(\dfrac{1 \text{ qt}}{946 \text{ mL}} \right) \left(\dfrac{1 \text{ gal}}{4 \text{ qt}} \right) = 7.48$ gal

3.31 $\dfrac{0.00133 \text{ g}}{\text{cm}^3}\left(\dfrac{1 \text{ kg}}{1000 \text{ g}}\right)\left(\dfrac{100 \text{ cm}}{1 \text{ m}}\right)^3 = \dfrac{0.00133 \text{ g}}{\text{cm}^3}\left(\dfrac{1 \text{ kg}}{1000 \text{ g}}\right)\left(\dfrac{100^3 \text{ cm}^3}{1^3 \text{ m}^3}\right) = 1.33 \text{ kg/m}^3$

3.32 $\dfrac{60.0 \text{ mi}}{\text{hr}}\left(\dfrac{1 \text{ hr}}{60 \text{ min}}\right)\left(\dfrac{1 \text{ min}}{60 \text{ sec}}\right)\left(\dfrac{5280 \text{ ft}}{1 \text{ mi}}\right)\left(\dfrac{12 \text{ in.}}{1 \text{ ft}}\right)\left(\dfrac{2.54 \text{ cm}}{1 \text{ in.}}\right)\left(\dfrac{1 \text{ m}}{100 \text{ cm}}\right) = 26.8 \text{ m/sec}$

3.33 $\dfrac{0.163 \text{ lb}}{\text{in.}^3}\left(\dfrac{1 \text{ in.}}{2.54 \text{ cm}}\right)^3\left(\dfrac{454 \text{ g}}{1 \text{ lb}}\right) = \dfrac{0.163 \text{ lb}}{\text{in.}^3}\left(\dfrac{1^3 \text{ in.}^3}{2.54^3 \text{ cm}^3}\right)\left(\dfrac{454 \text{ g}}{1 \text{ lb}}\right) = 4.52 \text{ g/cm}^3$

3.34 (a) An *exact number* has no uncertainty; for example, a dozen eggs is exactly twelve eggs. A *measured number* has some uncertainty; for example, the measurement of a person's height requires estimation of the last measured digit.

     (b) When carrying out a measurement, the last digit must be estimated. The degree to which that digit may be read represents its uncertainty.

     (c) *Precision* represents the agreement between repeated measurements of the same thing. *Accuracy* represents the agreement between a measurement and the correct value.

3.35 (a) Chemist 2 (b) Chemist 3 (c) Chemist 1

3.36 (a) 3 (b) 4 (c) 3 (d) 3 (e) 3 (f) 7 (g) 4 (h) 2 (i) 2 (j) 4

3.37 (a) 47.7 (b) 0.00821 (c) 20.0 (d) 17,500 (e) 6.56×10^{-4}

3.38 (a) 68.2 (b) 86 (c) 870 (d) 114 (e) 0.041 (f) 0.534 (g) 110

3.39   9

3.40 (a) $1.5 \text{ day}\left(\dfrac{24 \text{ hr}}{1 \text{ day}}\right) = 36 \text{ hr}$

     (b) $0.110 \text{ hr}\left(\dfrac{60 \text{ min}}{1 \text{ hr}}\right)\left(\dfrac{60 \text{ sec}}{1 \text{ min}}\right) = 396 \text{ sec}$

     (c) $21.0 \text{ day}\left(\dfrac{24 \text{ hr}}{1 \text{ day}}\right)\left(\dfrac{60 \text{ min}}{1 \text{ hr}}\right)\left(\dfrac{60 \text{ sec}}{1 \text{ min}}\right) = 1.81 \times 10^6 \text{ sec}$

     (d) $15.0 \text{ min}\left(\dfrac{1 \text{ hr}}{60 \text{ min}}\right)\left(\dfrac{1 \text{ day}}{24 \text{ hr}}\right)\left(\dfrac{1 \text{ yr}}{365 \text{ day}}\right) = 2.85 \times 10^{-5} \text{ yr}$

3.41 (a) $42 \text{ pottle}\left(\dfrac{1 \text{ puncheon}}{140 \text{ pottle}}\right) = 0.30 \text{ puncheon}$

3.41 (continued)

(b) $8.00 \text{ tun} \left(\dfrac{504 \text{ pottle}}{1 \text{ tun}} \right) \left(\dfrac{1 \text{ firkin}}{18 \text{ pottle}} \right) = 224 \text{ firkin}$

(c) $144 \text{ hogshead} \left(\dfrac{7 \text{ firkin}}{1 \text{ hogshead}} \right) \left(\dfrac{18 \text{ pottle}}{1 \text{ firkin}} \right) \left(\dfrac{1 \text{ tun}}{504 \text{ pottle}} \right) = 36.0 \text{ tun}$

3.42 $\$2.07 \left(\dfrac{100 \text{ cents}}{\$1.00} \right) \left(\dfrac{1 \text{ lb}}{69 \text{ cents}} \right) \left(\dfrac{6 \text{ apricots}}{1 \text{ lb}} \right) = 18 \text{ apricots}$

3.43 mega = 10^6; kilo = 1000; deci = 0.1; centi = 0.01; milli = 0.001; micro = 10^{-6}; nano = 10^{-9}.

3.44 meter; cubic meter; kilogram; Kelvin; Joule; pascal.

3.45 (a) $14.9 \text{ in.} \left(\dfrac{2.54 \text{ cm}}{1 \text{ in.}} \right) = 37.8 \text{ cm}$

(b) $0.427 \text{ in.} \left(\dfrac{2.54 \text{ cm}}{1 \text{ in.}} \right) \left(\dfrac{10 \text{ mm}}{1 \text{ cm}} \right) = 10.8 \text{ mm}$

(c) $31.0 \text{ ft} \left(\dfrac{12 \text{ in.}}{1 \text{ ft}} \right) \left(\dfrac{2.54 \text{ cm}}{1 \text{ in.}} \right) \left(\dfrac{1 \text{ m}}{100 \text{ cm}} \right) = 9.45 \text{ m}$

(d) $1.25 \text{ yd} \left(\dfrac{3 \text{ ft}}{1 \text{ yd}} \right) \left(\dfrac{12 \text{ in.}}{1 \text{ ft}} \right) \left(\dfrac{2.54 \text{ cm}}{1 \text{ in.}} \right) \left(\dfrac{1 \text{ m}}{100 \text{ cm}} \right) = 1.14 \text{ m}$

(e) $246 \text{ mi} \left(\dfrac{5280 \text{ ft}}{1 \text{ mi}} \right) \left(\dfrac{12 \text{ in.}}{1 \text{ ft}} \right) \left(\dfrac{2.54 \text{ cm}}{1 \text{ in.}} \right) \left(\dfrac{1 \text{ m}}{100 \text{ cm}} \right) \left(\dfrac{1 \text{ km}}{1000 \text{ m}} \right) = 396 \text{ km}$

(f) $4.52 \text{ in.} \left(\dfrac{2.54 \text{ cm}}{1 \text{ in.}} \right) \left(\dfrac{1 \text{ m}}{100 \text{ cm}} \right) \left(\dfrac{10^9 \text{ nm}}{1 \text{ m}} \right) = 1.15 \times 10^8 \text{ nm}$

3.46 Substitute your height in inches for x in the following setup:

$x \text{ in.} \left(\dfrac{2.54 \text{ cm}}{1 \text{ in.}} \right) \left(\dfrac{1 \text{ m}}{100 \text{ cm}} \right) = 0.0254x \text{ m}$

3.47 *Mass* is a measure of the quantity of matter. *Weight* is a measure of the force of gravity acting upon matter.

3.48 (a) 50.0 kg (b) 420 lb

3.49 balance, scale.

3.50 (a) $3.47 \text{ lb}\left(\dfrac{454 \text{ g}}{1 \text{ lb}}\right) = 1580 \text{ g} \ (1.58 \times 10^3 \text{ g})$

(b) $0.260 \text{ lb}\left(\dfrac{454 \text{ g}}{1 \text{ lb}}\right)\left(\dfrac{1000 \text{ mg}}{1 \text{ g}}\right) = 1.18 \times 10^5 \text{ mg}$

(c) $4.38 \text{ oz}\left(\dfrac{1 \text{ lb}}{16 \text{ oz}}\right)\left(\dfrac{454 \text{ g}}{1 \text{ lb}}\right) = 124 \text{ g}$

(d) $5.20 \text{ ton}\left(\dfrac{2000 \text{ lb}}{1 \text{ ton}}\right)\left(\dfrac{454 \text{ g}}{1 \text{ lb}}\right)\left(\dfrac{1 \text{ kg}}{1000 \text{ g}}\right) = 4720 \text{ kg} \ (4.72 \times 10^3 \text{ kg})$

(e) $3.50 \text{ ton}\left(\dfrac{2000 \text{ lb}}{1 \text{ ton}}\right)\left(\dfrac{454 \text{ g}}{1 \text{ lb}}\right)\left(\dfrac{1 \text{ Mg}}{10^6 \text{ g}}\right) = 3.18 \text{ Mg}$

(f) $0.653 \text{ mg}\left(\dfrac{1 \text{ g}}{10^3 \text{ mg}}\right)\left(\dfrac{10^6 \text{ μg}}{1 \text{ g}}\right) = 653 \text{ μg}$

3.51 Substitute your weight in pounds for y in the following setup:

$y \text{ lb}\left(\dfrac{454 \text{ g}}{1 \text{ lb}}\right)\left(\dfrac{1 \text{ kg}}{1000 \text{ g}}\right) = 0.454y \text{ kg}$

3.52 (a) $2.40 \text{ qt}\left(\dfrac{946 \text{ mL}}{1 \text{ qt}}\right)\left(\dfrac{1 \text{ L}}{1000 \text{ mL}}\right) = 2.27 \text{ L}$

(b) $43.0 \text{ L}\left(\dfrac{1000 \text{ mL}}{1 \text{ L}}\right)\left(\dfrac{1 \text{ qt}}{946 \text{ mL}}\right)\left(\dfrac{1 \text{ gal}}{4 \text{ qt}}\right) = 11.4 \text{ gal}$

(c) $12.0 \text{ oz}\left(\dfrac{1 \text{ qt}}{32 \text{ oz}}\right)\left(\dfrac{946 \text{ mL}}{1 \text{ qt}}\right)\left(\dfrac{1 \text{ L}}{1000 \text{ mL}}\right) = 0.355 \text{ L}$

3.53 $1 \text{ cup}\left(\dfrac{1 \text{ qt}}{4 \text{ cup}}\right)\left(\dfrac{946 \text{ mL}}{1 \text{ qt}}\right) = 236 \text{ mL}$

3.54 (a) *Density* is the mass per unit volume: $d = \dfrac{m}{V}$.

(b) Since most substances expand as they are heated, their densities must decrease.
(c) The molecules in a gas are much further apart than those in a liquid or solid.
(d) Gas bubbles are less dense than the liquid they are escaping from.
(e) Chloroform will sink, because it is more dense.

3.55 $d = \dfrac{18.3 \text{ g}}{25.0 \text{ mL}} = 0.732 \text{ g/mL}$

3.56 $35.0 \text{ mL} \left(\dfrac{13.6 \text{ g}}{1 \text{ mL}} \right) = 476 \text{ g}$

3.57 $1.32 \text{ L} \left(\dfrac{1000 \text{ mL}}{1 \text{ L}} \right) \left(\dfrac{0.879 \text{ g}}{1 \text{ mL}} \right) = 1160 \text{ g} \quad (1.16 \times 10^3 \text{ g})$

3.58 $654 \text{ g} \left(\dfrac{1 \text{ mL}}{1.48 \text{ g}} \right) = 442 \text{ mL}$

3.59 $2.44 \text{ kg} \left(\dfrac{1000 \text{ g}}{1 \text{ kg}} \right) \left(\dfrac{1 \text{ mL}}{0.714 \text{ g}} \right) = 3420 \text{ mL} \quad (\text{or } 3.42 \text{ L})$

3.60 m = 98.1 g; V = 46.0 mL - 35.0 mL = 11.0 mL

$d = \dfrac{m}{V} = \dfrac{98.1 \text{ g}}{11.0 \text{ mL}} = 8.92 \text{ g/mL}$

3.61 $V = (5.42 \text{ cm})(8.12 \text{ cm})(6.57 \text{ cm}) = 289 \text{ cm}^3$

$d = \dfrac{m}{V} = \dfrac{781 \text{ g}}{289 \text{ cm}^3} = 2.70 \text{ g/cm}^3 \quad (= 2.70 \text{ g/mL})$

Aluminum will sink, because its density is greater than that of water.

3.62 (a) *Specific gravity* is the ratio of the density of a substance compared to the density of water.

(b) The specific gravity of a substance is numerically equal to its density in grams per milliliter.

(c) If the specific gravity of a substance is greater than 1.00, it will sink, because its density must be greater than that of water.

3.63 (a) $\text{sp gr} = \dfrac{\text{density of lead}}{\text{density of water}} = \dfrac{708 \text{ lb / ft}^3}{62.4 \text{ lb / ft}^3} = 11.3$

(b) 11.3 g/mL

3.64 $T_F = 1.8T_C + 32$ $T_K = T_C + 273$

$-20.2 = 1.8T_C + 32$ $T_K = -29 + 273 = 244 \text{ K}$

$-52.2 = 1.8 \, T_C$

$-29°C = T_C$

3.65 $-179 = 1.8T_C + 32$ $T_K = -117 + 273 = 156 \text{ K}$

$-211 = 1.8 \, T_C$

$-117°C = T_C$

3.66 $98.6 = 1.8T_C + 32$ $T_K = 37 + 273 = 310 \text{ K}$

$66.6 = 1.8 \, T_C$

$37°C = T_C$

3.67 $T_F = 1.8(-40) + 32$
$T_F = -72 + 32$
$T_F = -40°F$

3.68 $T_F = 1.8(-273) + 32$
$T_F = -491 + 32$
$T_F = -459°F$

3.69 (a) 34 K: $34 = T_C + 273$
$T_C = -239°C$
$T_F = 1.8(-239) + 32$
$T_F = -398°F$

(b) 93 K: $93 = T_C + 273$
$T_C = -180°C$
$T_F = 1.8(-180) + 32$
$T_F = -292°F$

(c) $-196°C = 77$ K   Since 77 K < 93 K, Chu's ceramic will superconduct at this temperature.

(d) $-78°C = 195$ K   Since 195 K > 93 K, this temperature is too high for superconductivity to be observed.

(e) $-143°C = 130$ K This is 37 K higher than 93 K. Since the size of each Celsius degree is the same as that on the Kelvin scale, it is also 37 Celsius degrees higher.

3.70 (a) *Kinetic energy* is the energy of motion. *Potential energy* is energy that is due to the position of an object in a force field, such as a gravitational or magnetic field.

(b) Energy can be neither created nor destroyed.

(c) Heat, light, mechanical, electrical, chemical, and nuclear energy are forms of energy.

(d) One *calorie* is the quantity of heat required to raise the temperature of one gram of water by one degree Celsius.

3.71 (a) $11.2 \text{ cal}\left(\dfrac{4.184 \text{ J}}{1 \text{ cal}}\right) = 46.9$ J   (b) $7.31 \text{ J}\left(\dfrac{1 \text{ cal}}{4.184 \text{ J}}\right) = 1.75$ cal

(c) $893 \text{ cal}\left(\dfrac{4.184 \text{ J}}{1 \text{ cal}}\right)\left(\dfrac{1 \text{ kJ}}{1000 \text{ J}}\right) = 3.74$ kJ   (d) $6.53 \text{ kcal}\left(\dfrac{4.184 \text{ kJ}}{1 \text{ kcal}}\right) = 27.3$ kJ

3.72 *Heat* is a quantity of energy that may be transferred from one object to another. *Temperature* represents an average kinetic energy of matter.

3.73 (a) $43{,}560 \text{ ft}^2\left(\dfrac{1 \text{ yd}}{3 \text{ ft}}\right)^2 = 4840 \text{ yd}^2$

(b) $43{,}560 \text{ ft}^2\left(\dfrac{1 \text{ mi}}{5280 \text{ ft}}\right)^2 = 1.56 \times 10^{-3} \text{ mi}^2$

3.73 (continued)

(c) $43,560 \text{ ft}^2 \left(\dfrac{12 \text{ in.}}{1 \text{ ft}}\right)^2 \left(\dfrac{2.54 \text{ cm}}{1 \text{ in.}}\right)^2 \left(\dfrac{1 \text{ m}}{100 \text{ cm}}\right)^2 = 4050 \text{ m}^2 \quad (4.05 \times 10^3 \text{ m}^2)$

(d) $4.05 \times 10^3 \text{ m}^2 \left(\dfrac{1 \text{ km}}{1000 \text{ m}}\right)^2 = 4.05 \times 10^{-3} \text{ km}^2$

(e) $(106 \text{ ft})(82 \text{ ft})\left(\dfrac{1 \text{ acre}}{43,560 \text{ ft}^2}\right) = 0.20 \text{ acre}$

3.74 (a) $(50 \text{ m})\left(\dfrac{100 \text{ cm}}{1 \text{ m}}\right)\left(\dfrac{1 \text{ in.}}{2.54 \text{ cm}}\right)\left(\dfrac{1 \text{ ft}}{12 \text{ in.}}\right)(25.0 \text{ yd})\left(\dfrac{3 \text{ ft}}{1 \text{ yd}}\right)(5.00 \text{ ft})$

$= 6.15 \times 10^4 \text{ ft}^3$

(b) $6.15 \times 10^4 \text{ ft}^3 \left(\dfrac{12 \text{ in.}}{1 \text{ ft}}\right)^3 \left(\dfrac{2.54 \text{ cm}}{1 \text{ in.}}\right)^3 \left(\dfrac{1 \text{ m}}{100 \text{ cm}}\right)^3 = 1.74 \times 10^3 \text{ m}^3$

(c) $1.74 \times 10^3 \text{ m}^3 \left(\dfrac{100 \text{ cm}}{1 \text{ m}}\right)^3 \left(\dfrac{1 \text{ mL}}{1 \text{ cm}^3}\right)\left(\dfrac{1 \text{ L}}{1000 \text{ mL}}\right) = 1.74 \times 10^6 \text{ L}$

(d) $1.74 \times 10^6 \text{ L}\left(\dfrac{1000 \text{ mL}}{1 \text{ L}}\right)\left(\dfrac{1 \text{ qt}}{946 \text{ mL}}\right)\left(\dfrac{1 \text{ gal}}{4 \text{ qt}}\right) = 4.60 \times 10^5 \text{ gal}$

3.75 $V = (10.7 \text{ cm})(11.2 \text{ cm})(10.1 \text{ cm}) = 1210 \text{ cm}^3 = 1210 \text{ mL}$

$d = \dfrac{m}{V} = \dfrac{983 \text{ g}}{1210 \text{ mL}} = 0.812 \text{ g/mL}$

3.76 $5.00 \text{ lb}\left(\dfrac{454 \text{ g}}{1 \text{ lb}}\right)\left(\dfrac{1 \text{ cm}^3}{18.88 \text{ g}}\right) = 120 \text{ cm}^3 \quad (1.20 \times 10^2 \text{ cm}^3)$

3.77 (a) $\dfrac{3.00 \times 10^8 \text{ m}}{\text{sec}}\left(\dfrac{1 \text{ km}}{1000 \text{ m}}\right)\left(\dfrac{60 \text{ sec}}{1 \text{ min}}\right)\left(\dfrac{60 \text{ min}}{1 \text{ hr}}\right) = 1.08 \times 10^9 \text{ km/hr}$

(b) $\dfrac{1.08 \times 10^9 \text{ km}}{\text{hr}}\left(\dfrac{1000 \text{ m}}{1 \text{ km}}\right)\left(\dfrac{100 \text{ cm}}{1 \text{ m}}\right)\left(\dfrac{1 \text{ in.}}{2.54 \text{ cm}}\right)\left(\dfrac{1 \text{ ft}}{12 \text{ in.}}\right)\left(\dfrac{1 \text{ mi}}{5280}\right) = 6.71 \times 10^8 \text{ mi/hr}$

3.78 $\dfrac{1.00 \text{ g}}{\text{mL}}\left(\dfrac{1 \text{ lb}}{454 \text{ g}}\right)\left(\dfrac{1 \text{ mL}}{1 \text{ cm}^3}\right)\left(\dfrac{2.54 \text{ cm}}{1 \text{ in.}}\right)^3 \left(\dfrac{12 \text{ in.}}{1 \text{ ft}}\right)^3 = 62.4 \text{ lb/ft}^3$

3.79 Since more dense objects sink, the water at the bottom of the lake would be expected to be more dense than the water near the surface. The density of water increases from 0°C to 4°C. Thus, it follows that the colder (0°C), less dense water will be near the top of the lake, while the warmer (4°C), more dense water will sink to the bottom.

3.80 When a solid floats in a liquid, the volume of liquid displaced has a mass equal to the mass of the floating object. Thus, as the ice melts to become liquid water, the volume of liquid water produced exactly equals the volume of water displaced by the ice cube. In other words, when all of the ice is melted, the volume will come to the rim of the glass with no liquid overflow!

3.81 One way to do this problem is to convert 36.5°C and 37.5°C to Fahrenheit temperatures using the relationship: $T_F = 1.8T_C + 32$. This will give a range of 97.7°F to 99.5°F. However, the easier way to do this problem is to recognize that the size of a Celsius degree is 1.8 times as large as a Fahrenheit degree. Thus, 0.5 of a Celsius degree is 0.9 of a Fahrenheit degree. We can add and subtract 0.9°F to 98.6°F, giving the same result (97.7°F and 99.5°F).

3.82 This problem requires algebraic manipulation. We will take some liberties with significant figures to arrive at the answer. First we must combine the following two temperature relationships:

$$T_F = 1.8T_C + 32 \quad \text{and} \quad T_K = T_C + 273$$

Solving the second relationship for T_C and substituting the result into the first relationship gives the following:

$$T_C = T_K - 273$$
$$T_F = 1.8(T_K - 273) + 32$$

We will let T represent the temperature at which both scales have the same value:

$$T = T_F = T_K$$

Substituting this into our combined equation gives:

$$T = 1.8(T - 273) + 32$$
$$T = 1.8T - 491 + 32$$
$$-0.8T = -459$$
$$T = 574$$

Or:

$$574°F = 574 \text{ K}$$

If we check this answer, we will find that both temperatures equal 301°C.

3.83 There is less heat in an aluminum rod at 50°C than there is in a glass rod at the same temperature.

3.84 (a) divided by   (b) $\dfrac{91 \text{ cents}}{7 \text{ apples}} = 13$ cents/apple   (c) $\dfrac{91 \text{ cents}}{1.82 \text{ lb}} = 50$ cents/lb

(d) $\dfrac{1.82 \text{ lb}}{7 \text{ apples}} = 0.260$ lb/apple   (there are *exactly* 7 apples)

3.85 $\dfrac{454 \text{ g}}{16 \text{ oz}} = 28.4$ g/oz   (16 oz is an *exact* number)

3.86 (a) $\dfrac{225 \text{ mi}}{9.62 \text{ gal}} = 23.4$ mi/gal

(b) $\dfrac{9.62 \text{ gal}}{225 \text{ mi}} = 0.0428$ gal/mi

(c) $\dfrac{23.4 \text{ mi}}{1 \text{ gal}} \left(\dfrac{1 \text{ km}}{0.621 \text{ mi}} \right) \left(\dfrac{1 \text{ gal}}{3.785 \text{ L}} \right) = 9.96$ km/L

3.87 (a) $\dfrac{14.7 \text{ lb}}{1 \text{ in}^2} \left(\dfrac{1 \text{ in.}}{2.54 \text{ cm}} \right)^2 \left(\dfrac{454 \text{ g}}{1 \text{ lb}} \right) = 1030$ g/cm^2 = 1.03×10^3 g/cm^2

(b) $\dfrac{14.7 \text{ lb}}{1 \text{ in}^2} \left(\dfrac{1 \text{ in.}}{2.54 \text{ cm}} \right)^2 \left(\dfrac{100 \text{ cm}}{1 \text{ m}} \right)^2 \left(\dfrac{454 \text{ g}}{1 \text{ lb}} \right) \left(\dfrac{1 \text{ kg}}{1000 \text{ g}} \right) = 1.03 \times 10^4$ kg/m^2

3.88 $\dfrac{13.6 \text{ g}}{1 \text{ mL}} \left(\dfrac{1 \text{ lb}}{454 \text{ g}} \right) \left(\dfrac{1000 \text{ mL}}{1 \text{ L}} \right) \left(\dfrac{3.785 \text{ L}}{1 \text{ gal}} \right) = 113$ lb/gal

3.89 $50.0 \text{ gal} \left(\dfrac{4 \text{ qt}}{1 \text{ gal}} \right) \left(\dfrac{946 \text{ mL}}{1 \text{ qt}} \right) \left(\dfrac{1 \text{ cm}^3}{1 \text{ mL}} \right) \left(\dfrac{1 \text{ in.}}{2.54 \text{ cm}} \right)^3 \left(\dfrac{1 \text{ ft}}{12 \text{ in.}} \right)^3 = 6.68$ ft^3

3.90 $18.0 \text{ gal} \left(\dfrac{4 \text{ qt}}{1 \text{ gal}} \right) \left(\dfrac{946 \text{ mL}}{1 \text{ qt}} \right) \left(\dfrac{1 \text{ L}}{1000 \text{ mL}} \right) \left(\dfrac{32.9 \text{ cents}}{1 \text{ L}} \right) \left(\dfrac{\$1.00}{100 \text{ cents}} \right) = \$22.4$

(The cents column in 22.4 dollars is omitted because it is not a significant figure.)

3.91 $\$125 \left(\dfrac{1 \text{ gal}}{\$1.33} \right) \left(\dfrac{37.0 \text{ mi}}{1 \text{ gal}} \right) = 3480$ mi   (3.48×10^3 mi)

3.92 $35.0 \text{ L} \left(\dfrac{1000 \text{ mL}}{1 \text{ L}} \right) \left(\dfrac{1 \text{ qt}}{946 \text{ mL}} \right) \left(\dfrac{1 \text{ gal}}{4 \text{ qt}} \right) \left(\dfrac{42.3 \text{ mi}}{1 \text{ gal}} \right) = 391$ mi

3.93 $1 \text{ lb} \left(\dfrac{16 \text{ oz}}{1 \text{ lb}} \right) \left(\dfrac{1 \text{ cracker}}{0.100 \text{ oz}} \right) \left(\dfrac{1 \text{ box}}{80 \text{ cracker}} \right) \left(\dfrac{\$1.29}{1 \text{ box}} \right) = \$2.58$

3.94 $10.25 \text{ min} \left(\dfrac{60 \text{ sec}}{1 \text{ min}} \right) \left(\dfrac{1 \text{ car}}{8.2 \text{ sec}} \right) \left(\dfrac{24 \text{ ft}}{1 \text{ car}} \right) = 1800 \text{ ft} \ (1.8 \times 10^3 \text{ ft})$

Round-trip distance is 1800 ft. One-way distance is 900 ft (9.0×10^3 ft).

3.95 $\$5.67 \left(\dfrac{1 \text{ bag}}{\$1.89} \right) \left(\dfrac{10 \text{ lb}}{1 \text{ bag}} \right) = 30 \text{ lb}$

3.96 $1 \text{ cherry} \left(\dfrac{1 \text{ lb}}{58 \text{ cherry}} \right) \left(\dfrac{454 \text{ g}}{1 \text{ lb}} \right) = 7.8 \text{ g}$

3.97 $1 \text{ green pepper} \left(\dfrac{74 \text{ cents}}{2 \text{ green pepper}} \right) \left(\dfrac{1 \text{ lb}}{89 \text{ cents}} \right) \left(\dfrac{454 \text{ g}}{1 \text{ lb}} \right) = 190 \text{ g} \ (1.9 \times 10^2 \text{ g})$

CHAPTER 4

Quantitative Description of Matter

4.1 (a) C_2H_5 (b) CCl_4 (c) BH_3 (d) CH_2O (e) NH_2

4.2 (a) 83.8 (b) 85.5 (c) 238.0 (d) 102.9 (e) 195.1 (f) 79.9 (g) 40.1 (h) 27.0
(i) 126.9 (j) 31.0

4.3 (a) $2(23.0) + 12.0 + 3(16.0) = 106.0$ (b) $209.0 + 3(35.5) = 315.5$
(c) $39.1 + 54.9 + 4(16.0) = 158.0$ (d) $23.0 + 1.0 + 12.0 + 3(16.0) = 84.0$
(e) $2(12.0) + 5(1.0) + 35.5 = 64.5$ (f) $40.1 + 2(16.0) + 2(1.0) = 74.1$
(g) $2(27.0) + 3(32.1) + 12(16.0) = 342.3$

4.4 (a)
$$
\begin{aligned}
1 \times \text{Mn} &= 1 \times 54.9 = 54.9 \\
2 \times \text{O} &= 2 \times 16.0 = 32.0 \\
\hline
\text{MnO}_2 &= 86.9
\end{aligned}
$$

$$\%\text{Mn} = \frac{54.9}{86.9} \times 100\% = 63.2\%$$

$$\%\text{O} = \frac{32.0}{86.9} \times 100\% = 36.8\%$$

(b)
$$
\begin{aligned}
3 \times \text{C} &= 3 \times 12.0 = 36.0 \\
8 \times \text{F} &= 8 \times 19.0 = 152.0 \\
\hline
\text{C}_3\text{F}_8 &= 188.0
\end{aligned}
$$

$$\%\text{C} = \frac{36.0}{188.0} \times 100\% = 19.1\%$$

$$\%\text{F} = \frac{152.0}{188.0} \times 100\% = 80.9\%$$

(c)
$$
\begin{aligned}
1 \times \text{Cu} &= 1 \times 63.5 = 63.5 \\
2 \times \text{Br} &= 2 \times 79.9 = 159.8 \\
\hline
\text{CuBr}_2 &= 223.3
\end{aligned}
$$

$$\%\text{Cu} = \frac{63.5}{223.3} \times 100\% = 28.4\%$$

$$\%\text{Br} = \frac{159.8}{223.3} \times 100\% = 71.6\%$$

(d)
$$
\begin{aligned}
3 \times \text{H} &= 3 \times 1.0 = 3.0 \\
1 \times \text{P} &= 1 \times 31.0 = 31.0 \\
4 \times \text{O} &= 4 \times 16.0 = 64.0 \\
\hline
\text{H}_3\text{PO}_4 &= 98.0
\end{aligned}
$$

$$\%\text{H} = \frac{3.0}{98.0} \times 100\% = 3.1\%$$

$$\%\text{P} = \frac{31.0}{98.0} \times 100\% = 31.6\%$$

$$\%\text{O} = \frac{64.0}{98.0} \times 100\% = 65.3\%$$

(e)
$$
\begin{aligned}
1 \times \text{C} &= 1 \times 12.0 = 12.0 \\
2 \times \text{H} &= 2 \times 1.0 = 2.0 \\
1 \times \text{Br} &= 1 \times 79.9 = 79.9 \\
1 \times \text{F} &= 1 \times 19.0 = 19.0 \\
\hline
\text{CH}_2\text{BrF} &= 112.9
\end{aligned}
$$

$$\%\text{C} = \frac{12.0}{112.9} \times 100\% = 10.6\%$$

$$\%\text{H} = \frac{2.0}{112.9} \times 100\% = 1.8\%$$

$$\%\text{Br} = \frac{79.9}{112.9} \times 100\% = 70.8\%$$

$$\%\text{F} = \frac{19.0}{112.9} \times 100\% = 16.8\%$$

(f)
$$1 \times Zn = 1 \times 65.4 = 65.4$$
$$2 \times N = 2 \times 14.0 = 28.0$$
$$\underline{6 \times O = 6 \times 16.0 = 96.0}$$
$$Zn(NO_3)_2 = 189.4$$

$$\%Zn = \frac{65.4}{189.9} \times 100\% = 34.5\%$$
$$\%N = \frac{28.0}{189.4} \times 100\% = 14.8\%$$
$$\%O = \frac{96.0}{189.4} \times 100\% = 50.7\%$$

(g)
$$1 \times Mg = 1 \times 24.3 = 24.3$$
$$2 \times C = 2 \times 12.0 = 24.0$$
$$\underline{2 \times N = 2 \times 14.0 = 28.0}$$
$$Mg(CN)_2 = 76.3$$

$$\%Mg = \frac{24.3}{76.3} \times 100\% = 31.8\%$$
$$\%C = \frac{24.0}{76.3} \times 100\% = 31.5\%$$
$$\%N = \frac{28.0}{76.3} \times 100\% = 36.7\%$$

4.5 (a)
$$1 \times Na = 1 \times 23.0 = 23.0$$
$$\underline{1 \times Cl = 1 \times 35.5 = 35.5}$$
$$NaCl = 58.5$$

$$3.50 \text{ g NaCl} \left(\frac{23.0 \text{ g Na}}{58.5 \text{ g NaCl}} \right) = 1.38 \text{ g Na}$$

(b)
$$1 \times Ag = 1 \times 107.9 = 107.9$$
$$1 \times N = 1 \times 14.0 = 14.0$$
$$\underline{3 \times O = 3 \times 16.0 = 48.0}$$
$$AgNO_3 = 169.9$$

$$5.65 \text{ g AgNO}_3 \left(\frac{107.9 \text{ g Ag}}{169.9 \text{ g AgNO}_3} \right)$$
$$= 3.59 \text{ g Ag}$$

(c)
$$2 \times N = 2 \times 14.0 = 28.0$$
$$4 \times H = 4 \times 1.0 = 4.0$$
$$\underline{3 \times O = 3 \times 16.0 = 48.0}$$
$$NH_4NO_3 = 80.0$$

$$454 \text{ g NH}_4NO_3 \left(\frac{28.0 \text{ g N}}{80.0 \text{ g NH}_4NO_3} \right)$$
$$= 159 \text{ g N}$$

(d)
$$1 \times Fe = 1 \times 55.8 = 55.8$$
$$1 \times S = 1 \times 32.1 = 32.1$$
$$\underline{4 \times O = 4 \times 16.0 = 64.0}$$
$$FeSO_4 = 151.9$$

$$125 \text{ mg FeSO}_4 \left(\frac{55.8 \text{ mg Fe}}{151.9 \text{ mg FeSO}_4} \right)$$
$$= 45.9 \text{ mg Fe}$$

4.6 (a) $6.00 \text{ g C} \left(\frac{1 \text{ mol C}}{12.0 \text{ g C}} \right) = 0.500 \text{ mol C}$

(b) $4.19 \text{ g Kr} \left(\frac{1 \text{ mol Kr}}{83.8 \text{ g Kr}} \right) = 0.0500 \text{ mol Kr}$

(c) $47.6 \text{ g U} \left(\frac{1 \text{ mol U}}{238.0 \text{ g U}} \right) = 0.200 \text{ mol U}$

4.6 (continued)

(d) $8.37 \text{ g Fe} \left(\dfrac{1 \text{ mol Fe}}{55.8 \text{ g Fe}} \right) = 0.150 \text{ mol Fe}$

(e) $8.82 \text{ g Ca} \left(\dfrac{1 \text{ mol Ca}}{40.1 \text{ g Ca}} \right) = 0.220 \text{ mol Ca}$

(f) $316 \text{ g Mg} \left(\dfrac{1 \text{ mol Mg}}{24.3 \text{ g Mg}} \right) = 13.0 \text{ mol Mg}$

4.7 (a) $2.50 \text{ mol Ne} \left(\dfrac{20.2 \text{ g Ne}}{1 \text{ mol Ne}} \right) = 50.5 \text{ g Ne}$

(b) $0.452 \text{ mol Na} \left(\dfrac{23.0 \text{ g Na}}{1 \text{ mol Na}} \right) = 10.4 \text{ g Na}$

(c) $0.125 \text{ mol K} \left(\dfrac{39.1 \text{ g K}}{1 \text{ mol K}} \right) = 4.89 \text{ g K}$

(d) $0.742 \text{ mol Hg} \left(\dfrac{200.6 \text{ g Hg}}{1 \text{ mol Hg}} \right) = 149 \text{ g Hg}$

(e) $0.0531 \text{ mol Au} \left(\dfrac{197.0 \text{ g Au}}{1 \text{ mol Au}} \right) = 10.5 \text{ g Au}$

4.8 (a) $0.25 \text{ mol He} \left(\dfrac{6.02 \times 10^{23} \text{ atoms He}}{1 \text{ mol He}} \right) = 1.5 \times 10^{23} \text{ atoms He}$

(b) $3.50 \text{ mol Fe} \left(\dfrac{6.02 \times 10^{23} \text{ atoms Fe}}{1 \text{ mol Fe}} \right) = 2.11 \times 10^{24} \text{ atoms Fe}$

(c) $0.0155 \text{ mol Au} \left(\dfrac{6.02 \times 10^{23} \text{ atoms Au}}{1 \text{ mol Au}} \right) = 9.33 \times 10^{21} \text{ atoms Au}$

(d) $6.35 \text{ g Cu} \left(\dfrac{1 \text{ mol Cu}}{63.5 \text{ g Cu}} \right) \left(\dfrac{6.02 \times 10^{23} \text{ atoms Cu}}{1 \text{ mol Cu}} \right) = 6.02 \times 10^{22} \text{ atoms Cu}$

(e) $253 \text{ g Na} \left(\dfrac{1 \text{ mol Na}}{23.0 \text{ g Na}} \right) \left(\dfrac{6.02 \times 10^{23} \text{ atoms Na}}{1 \text{ mol Na}} \right) = 6.62 \times 10^{24} \text{ atoms Na}$

(f) $10.0 \text{ g S} \left(\dfrac{1 \text{ mol S}}{32.1 \text{ g S}} \right) \left(\dfrac{6.02 \times 10^{23} \text{ atoms S}}{1 \text{ mol S}} \right) = 1.88 \times 10^{23} \text{ atoms S}$

4.9 (a) $1.56 \text{ g CO} \left(\dfrac{1 \text{ mol CO}}{28.0 \text{ g CO}} \right) = 0.0557 \text{ mol CO}$

(b) $15.0 \text{ g CO}_2 \left(\dfrac{1 \text{ mol CO}_2}{44.0 \text{ g CO}_2} \right) = 0.341 \text{ mol CO}_2$

(c) $2.50 \text{ g NO}_2 \left(\dfrac{1 \text{ mol NO}_2}{46.0 \text{ g NO}_2} \right) = 0.0543 \text{ mol NO}_2$

(d) $12.5 \text{ g KBr} \left(\dfrac{1 \text{ mol KBr}}{119.0 \text{ g KBr}} \right) = 0.105 \text{ mol KBr}$

(e) $20.0 \text{ g K}_2\text{CO}_3 \left(\dfrac{1 \text{ mol K}_2\text{CO}_3}{138.2 \text{ g K}_2\text{CO}_3} \right) = 0.145 \text{ mol K}_2\text{CO}_3$

(f) $12.1 \text{ g SO}_3 \left(\dfrac{1 \text{ mol SO}_3}{80.1 \text{ g SO}_3} \right) = 0.151 \text{ mol SO}_3$

4.10 (a) $2.50 \text{ mol CH}_4 \left(\dfrac{16.0 \text{ g CH}_4}{1 \text{ mol CH}_4} \right) = 40.0 \text{ g CH}_4$

(b) $0.550 \text{ mol KCl} \left(\dfrac{74.6 \text{ g KCl}}{1 \text{ mol KCl}} \right) = 41.0 \text{ g KCl}$

(c) $0.315 \text{ mol NaOH} \left(\dfrac{40.0 \text{ g NaOH}}{1 \text{ mol NaOH}} \right) = 12.6 \text{ g NaOH}$

(d) $0.114 \text{ mol H}_2\text{SO}_4 \left(\dfrac{98.1 \text{ g H}_2\text{SO}_4}{1 \text{ mol H}_2\text{SO}_4} \right) = 11.2 \text{ g H}_2\text{SO}_4$

4.11 (a) $5.0 \text{ g H}_2 \left(\dfrac{1 \text{ mol H}_2}{2.0 \text{ g H}_2} \right) = 2.5 \text{ mol H}_2$

(b) $20.0 \text{ g Cl}_2 \left(\dfrac{1 \text{ mol Cl}_2}{71.0 \text{ g Cl}_2} \right) = 0.282 \text{ mol Cl}_2$

(c) $11.3 \text{ g I} \left(\dfrac{1 \text{ mol I}}{126.9 \text{ g I}} \right) = 0.0890 \text{ mol I}$

(d) $11.3 \text{ g I}_2 \left(\dfrac{1 \text{ mol I}_2}{253.8 \text{ g I}_2} \right) = 0.0445 \text{ mol I}_2$

(e) $1.40 \text{ g Ar} \left(\dfrac{1 \text{ mol Ar}}{39.9 \text{ g Ar}} \right) = 0.0351 \text{ mol Ar}$

4.11 (continued)

(f) $17.0 \text{ g F}_2 \left(\dfrac{1 \text{ mol F}_2}{38.0 \text{ g F}_2} \right) = 0.447 \text{ mol F}_2$

4.12 (a) $4.75 \text{ mol NH}_3 \left(\dfrac{6.02 \times 10^{23} \text{ molecules NH}_3}{1 \text{ mol NH}_3} \right) = 2.86 \times 10^{24} \text{ molecules NH}_3$

(b) $0.450 \text{ mol CCl}_4 \left(\dfrac{6.02 \times 10^{23} \text{ molecules CCl}_4}{1 \text{ mol CCl}_4} \right) = 2.71 \times 10^{23} \text{ molecules CCl}_4$

(c) $0.0135 \text{ mol F}_2 \left(\dfrac{6.02 \times 10^{23} \text{ molecules F}_2}{1 \text{ mol F}_2} \right) = 8.13 \times 10^{21} \text{ molecules F}_2$

(d) $14.7 \text{ g C}_3\text{H}_6\text{O} \left(\dfrac{1 \text{ mol C}_3\text{H}_6\text{O}}{58.0 \text{ g C}_3\text{H}_6\text{O}} \right) \left(\dfrac{6.02 \times 10^{23} \text{ molecules C}_3\text{H}_6\text{O}}{1 \text{ mol C}_3\text{H}_6\text{O}} \right)$

  $= 1.53 \times 10^{23} \text{ molecules C}_3\text{H}_6\text{O}$

(e) $8.55 \text{ g N}_2 \left(\dfrac{1 \text{ mol N}_2}{28.0 \text{ g N}_2} \right) \left(\dfrac{6.02 \times 10^{23} \text{ molecules N}_2}{1 \text{ mol N}_2} \right) = 1.84 \times 10^{23} \text{ molecules N}_2$

4.13 (a) $52.4 \text{ g K} \left(\dfrac{1 \text{ mol K}}{39.1 \text{ g K}} \right) = 1.34 \text{ mol K};$ $\dfrac{1.34}{1.34} = 1$

  $47.6 \text{ g Cl} \left(\dfrac{1 \text{ mol Cl}}{35.5 \text{ g Cl}} \right) = 1.34 \text{ mol Cl};$ $\dfrac{1.34}{1.34} = 1$

  empirical formula: KCl

(b) $16.2 \text{ g Na} \left(\dfrac{1 \text{ mol Na}}{23.0 \text{ g Na}} \right) = 0.704 \text{ mol Na};$ $\dfrac{0.704}{0.703} \approx 1$

  $38.6 \text{ g Mn} \left(\dfrac{1 \text{ mol Mn}}{54.9 \text{ g Mn}} \right) = 0.703 \text{ mol Mn};$ $\dfrac{0.703}{0.703} = 1$

  $45.2 \text{ g O} \left(\dfrac{1 \text{ mol O}}{16.0 \text{ g O}} \right) = 2.82 \text{ mol O};$ $\dfrac{2.82}{0.703} \approx 4$

  empirical formula: NaMnO_4

4.13 (continued)

(c) $3.1 \text{ g H} \left(\dfrac{1 \text{ mol H}}{1.0 \text{ g H}} \right) = 3.1 \text{ mol H};$ $\qquad$ $\dfrac{3.1}{1.02} \approx 3$

$31.5 \text{ g P} \left(\dfrac{1 \text{ mol P}}{31.0 \text{ g P}} \right) = 1.02 \text{ mol P};$ $\qquad$ $\dfrac{1.02}{1.02} = 1$

$65.4 \text{ g O} \left(\dfrac{1 \text{ mol O}}{16.0 \text{ g O}} \right) = 4.09 \text{ mol O};$ $\qquad$ $\dfrac{4.09}{1.02} \approx 4$

empirical formula: H_3PO_4

(d) $56.6 \text{ g K} \left(\dfrac{1 \text{ mol K}}{39.1 \text{ g K}} \right) = 1.45 \text{ mol K};$ $\qquad$ $\dfrac{1.45}{0.72} \approx 2$

$8.7 \text{ g C} \left(\dfrac{1 \text{ mol C}}{12.0 \text{ g C}} \right) = 0.72 \text{ mol C};$ $\qquad$ $\dfrac{0.72}{0.72} = 1$

$34.7 \text{ g O} \left(\dfrac{1 \text{ mol O}}{16.0 \text{ g O}} \right) = 2.17 \text{ mol O};$ $\qquad$ $\dfrac{2.17}{0.72} \approx 3$

empirical formula: K_2CO_3

(e) $10.04 \text{ g C} \left(\dfrac{1 \text{ mol C}}{12.0 \text{ g C}} \right) = 0.837 \text{ mol C};$ $\qquad$ $\dfrac{0.837}{0.837} = 1$

$0.84 \text{ g H} \left(\dfrac{1 \text{ mol H}}{1.0 \text{ g H}} \right) = 0.84 \text{ mol H};$ $\qquad$ $\dfrac{0.84}{0.837} \approx 1$

$89.12 \text{ g Cl} \left(\dfrac{1 \text{ mol Cl}}{35.5 \text{ g Cl}} \right) = 2.51 \text{ mol Cl};$ $\qquad$ $\dfrac{2.51}{0.837} \approx 3$

empirical formula: $CHCl_3$

4.14 $1.67 \text{ g Ca} \left(\dfrac{1 \text{ mol Ca}}{40.1 \text{ g Ca}} \right) = 0.0416 \text{ mol Ca};$ $\qquad$ $\dfrac{0.0416}{0.0416} = 1$

$2.96 \text{ g Cl} \left(\dfrac{1 \text{ mol Cl}}{35.5 \text{ g Cl}} \right) = 0.0834 \text{ mol Cl};$ $\qquad$ $\dfrac{0.0834}{0.0416} \approx 2$

empirical formula: $CaCl_2$

4.15 $0.432 \text{ g C} \left(\dfrac{1 \text{ mol C}}{12.0 \text{ g C}} \right) = 0.0360 \text{ mol C};$ $\qquad$ $\dfrac{0.0360}{0.0180} = 2$

$0.090 \text{ g H} \left(\dfrac{1 \text{ mol H}}{1.0 \text{ g H}} \right) = 0.090 \text{ mol H};$ $\qquad$ $\dfrac{0.090}{0.0180} = 5$

4.15 (continued)

$$0.342 \text{ g F} \left(\frac{1 \text{ mol F}}{19.0 \text{ g F}} \right) = 0.0180 \text{ mol F}; \qquad \frac{0.0180}{0.0180} = 1$$

empirical formula: C_2H_5F

4.16 (a) $\dfrac{165 \text{ g}}{1.25 \text{ mol}} = 132 \text{ g/mol}$

(b) $\dfrac{14.5}{0.320 \text{ mol}} = 45.3 \text{ g/mol}$

(c) $\dfrac{2.35}{0.0155 \text{ mol}} = 152 \text{ g/mol}$

4.17 (a) Empirical formula mass for CH: $12.0 + 1.0 = 13.0$

$\dfrac{52.0}{13.0} = 4;$ molecular formula: C_4H_4

(b) Empirical formula mass for CH: $12.0 + 1.0 = 13.0$

$\dfrac{78.0}{13.0} = 6;$ molecular formula: C_6H_6

(c) Empirical formula mass for C_2H_3F: $2(12.0) + 3(1.0) + 19.0 = 46.0$

$\dfrac{92.0}{46.0} = 2;$ molecular formula: $C_4H_6F_2$

(d) Empirical formula mass for NaO: $23.0 + 16.0 = 39.0$

$\dfrac{78.0}{39.0} = 2;$ molecular formula: Na_2O_2

(e) Empirical formula mass for C_5H_5N: $5(12.0) + 5(1.0) + 14.0 = 79.0$

$\dfrac{79.0}{79.0} = 1;$ molecular formula: C_5H_5N

(f) Empirical formula mass for C_2HNO_2: $2(12.0) + 1.0 + 14.0 + 2(16.0) = 71.0$

$\dfrac{213.0}{71.0} = 3;$ molecular formula: $C_6H_3N_3O_6$

4.18 A molecular formula tells how many atoms of each element are present in a molecule of a substance, whereas an empirical formula only gives the simplest whole-number ratio of atoms present.

4.19 (a) HO   (b) C_2H_6O (c) $C_2H_3O_3$ (d) C_3H_3O (e) CH_2O

4.20 An *atomic mass* is the relative mass of an atom of an element as compared to some standard. (In Chapter 5 we define atomic mass of an element more precisely as the average mass of its naturally occuring isotopes, as compared to a standard of carbon-12 equal to exactly 12 u.) A *relative mass* is a ratio of the mass of an object compared to some other object.

4.21 (a) 6.9   (b) 20.2   (c) 10.8   (d) 32.1   (e) 35.5   (f) 39.9   (g) 79.0   (h) 55.8
(i) 63.5   (j) 65.4

4.22 (a)

$1 \times$ Na	=	1×23.0	=	23.0
$1 \times$ O	=	1×16.0	=	16.0
$1 \times$ Cl	=	1×35.5	=	35.5
		NaOCl	=	74.5

(b)

$4 \times$ C	=	4×12.0	=	48.0
$10 \times$ H	=	10×1.0	=	10.0
		C_4H_{10}	=	58.0

(c)

$1 \times$ H	=	1×1.0	=	1.0
$1 \times$ N	=	1×14.0	=	14.0
$3 \times$ O	=	3×16.0	=	48.0
		HNO_3	=	63.0

(d)

$1 \times$ Al	=	1×27.0	=	27.0
$3 \times$ O	=	3×16.0	=	48.0
$3 \times$ H	=	3×1.0	=	3.0
		$Al(OH)_3$	=	78.0

(e)

$3 \times$ Zn	=	3×65.4	=	196.2
$2 \times$ P	=	2×31.0	=	62.0
$8 \times$ O	=	8×16.0	=	128.0
		$Zn_3(PO_4)_2$	=	386.2

4.23 (a)

$2 \times$ N	=	2×14.0	=	28.0
$1 \times$ O	=	1×16.0	=	16.0
		N_2O	=	44.0

$$\% \text{ N} = \frac{28.0}{44.0} \times 100\% = 63.6\%$$

$$\% \text{ O} = \frac{16.0}{44.0} \times 100\% = 36.4\%$$

(b)

$2 \times$ C	=	2×12.0	=	24.0
$6 \times$ H	=	6×1.0	=	6.0
$1 \times$ O	=	1×16.0	=	16.0
		C_2H_6O	=	46.0

$$\% \text{ C} = \frac{24.0}{46.0} \times 100\% = 52.2\%$$

$$\% \text{ H} = \frac{6.0}{46.0} \times 100\% = 13.0\%$$

$$\% \text{ O} = \frac{16.0}{46.0} \times 100\% = 34.8\%$$

(c)

$14 \times$ C	=	14×12.0	=	168.0
$9 \times$ H	=	9×1.0	=	9.0
$5 \times$ Cl	=	5×35.5	=	177.5
		$C_{14}H_9Cl_5$	=	354.5

$$\% \text{ C} = \frac{168.0}{354.5} \times 100\% = 47.4\%$$

$$\% \text{ H} = \frac{9.0}{354.5} \times 100\% = 2.5\%$$

$$\% \text{ Cl} = \frac{177.5}{354.5} \times 100\% = 50.1\%$$

(d)

$7 \times$ C	=	7×12.0	=	84.0
$7 \times$ H	=	7×1.0	=	7.0
$2 \times$ O	=	2×16.0	=	32.0
$1 \times$ N	=	1×14.0	=	14.0
		$C_7H_7O_2N$	=	137.0

$$\% \text{ C} = \frac{84.0}{137.0} \times 100\% = 61.3\%$$

$$\% \text{ H} = \frac{7.0}{137.0} \times 100\% = 5.1\%$$

$$\% \text{ O} = \frac{32.0}{137.0} \times 100\% = 23.4\%$$

$$\% \text{ N} = \frac{14.0}{137.0} \times 100\% = 10.2\%$$

4.23 (continued)

(e)
$$9 \times C = 9 \times 12.0 = 108.0$$
$$13 \times H = 13 \times 1.0 = 13.0$$
$$3 \times O = 3 \times 16.0 = 48.0$$
$$\underline{1 \times N = 1 \times 14.0 = 14.0}$$
$$C_9H_{13}O_3N = 183.0$$

$$\% C = \frac{108.0}{183.0} \times 100\% = 59.0\%$$
$$\% H = \frac{13.0}{183.0} \times 100\% = 7.1\%$$
$$\% O = \frac{48.0}{183.0} \times 100\% = 26.2\%$$
$$\% N = \frac{14.0}{183.0} \times 100\% = 7.7\%$$

(f)
$$1 \times K = 1 \times 39.1 = 39.1$$
$$4 \times C = 4 \times 12.0 = 48.0$$
$$5 \times H = 5 \times 1.0 = 5.0$$
$$\underline{6 \times O = 6 \times 16.0 = 96.0}$$
$$KC_4H_5O_6 = 188.1$$

$$\% K = \frac{39.1}{188.1} \times 100\% = 20.8\%$$
$$\% C = \frac{48.0}{188.1} \times 100\% = 25.5\%$$
$$\% H = \frac{5.0}{188.1} \times 100\% = 2.7\%$$
$$\% O = \frac{96.0}{188.1} \times 100\% = 51.0\%$$

(g)
$$1 \times Na = 1 \times 23.0 = 23.0$$
$$5 \times C = 5 \times 12.0 = 60.0$$
$$8 \times H = 8 \times 1.0 = 8.0$$
$$4 \times O = 4 \times 16.0 = 64.0$$
$$\underline{1 \times N = 1 \times 14.0 = 14.0}$$
$$NaC_5H_8O_4N = 169.0$$

$$\% Na = \frac{23.0}{169.0} \times 100\% = 13.6\%$$
$$\% C = \frac{60.0}{169.0} \times 100\% = 35.5\%$$
$$\% H = \frac{8.0}{169.0} \times 100\% = 4.7\%$$
$$\% O = \frac{64.0}{169.0} \times 100\% = 37.9\%$$
$$\% N = \frac{14.0}{169.0} \times 100\% = 8.3\%*$$
*(8.28% rounded off to nearest 0.1%)

(h)
$$1 \times Mg = 1 \times 24.3 = 24.3$$
$$4 \times C = 4 \times 12.0 = 48.0$$
$$6 \times H = 6 \times 1.0 = 6.0$$
$$\underline{4 \times O = 4 \times 16.0 = 64.0}$$
$$Mg(C_2H_3O_2)_2 = 142.3$$

$$\% Mg = \frac{24.3}{142.3} \times 100\% = 17.1\%$$
$$\% C = \frac{48.0}{142.3} \times 100\% = 33.7\%$$
$$\% H = \frac{6.0}{142.3} \times 100\% = 4.2\%$$
$$\% O = \frac{64.0}{142.3} \times 100\% = 45.0\%$$

4.23 (continued)

(i)

$$
\begin{array}{lllll}
2 \times N & = & 2 \times 14.0 & = & 28.0 \\
8 \times H & = & 8 \times 1.0 & = & 8.0 \\
1 \times S & = & 1 \times 32.1 & = & 32.1 \\
4 \times O & = & \underline{4 \times 16.0} & = & \underline{64.0} \\
& & N_2H_8SO_4 & = & 132.1
\end{array}
$$

$$\% \ N = \frac{28.0}{132.1} \times 100\% = \quad 21.2\%$$

$$\% \ H = \frac{8.0}{132.1} \times 100\% = \quad 6.1\%$$

$$\% \ S = \frac{32.1}{132.1} \times 100\% = \quad 24.3\%$$

$$\% \ O = \frac{64.0}{132.1} \times 100\% = \quad 48.4\%$$

4.24 (a)

$$
\begin{array}{lllll}
1 \times K & = & 1 \times 39.1 & = & 39.1 \\
1 \times I & = & \underline{1 \times 126.9} & = & \underline{126.9} \\
& & KI & = & 166.0
\end{array}
$$

$$1.25 \text{ g KI} \left(\frac{126.9 \text{ g I}}{166.0 \text{ g KI}} \right) = 0.956 \text{ g I}$$

(b)

$$
\begin{array}{lllll}
1 \times Ca & = & 1 \times 40.1 & = & 40.1 \\
2 \times C & = & 2 \times 12.0 & = & 24.0 \\
4 \times O & = & \underline{4 \times 16.0} & = & \underline{64.0} \\
& & CaC_2O_4 & = & 128.1
\end{array}
$$

$$0.750 \text{ g } CaC_2O_4 \left(\frac{40.1 \text{ g Ca}}{128.1 \text{ g } CaC_2O_4} \right)$$
$$= 0.235 \text{ g Ca}$$

(c)

$$
\begin{array}{lllll}
3 \times Na & = & 3 \times 23.0 & = & 69.0 \\
1 \times P & = & 1 \times 31.0 & = & 31.0 \\
4 \times O & = & \underline{4 \times 16.0} & = & \underline{64.0} \\
& & Na_3PO_4 & = & 164.0
\end{array}
$$

$$3.85 \text{ g } Na_3PO_4 \left(\frac{31.0 \text{ g P}}{164.0 \text{ g } Na_3PO_4} \right)$$
$$= 0.728 \text{ g P}$$

(d)

$$
\begin{array}{lllll}
1 \times Mg & = & 1 \times 24.3 & = & 24.3 \\
2 \times O & = & 2 \times 16.0 & = & 32.0 \\
2 \times H & = & \underline{2 \times 1.0} & = & \underline{2.0} \\
& & Mg(OH)_2 & = & 58.3
\end{array}
$$

$$275 \text{ mg } Mg(OH)_2 \left(\frac{24.3 \text{ mg Mg}}{58.3 \text{ mg } Mg(OH)_2} \right)$$
$$= 115 \text{ mg } Mg(OH)_2$$

4.25 (a) One mole equals Avogadro's number of particles: 6.02×10^{23}.

(b) One mole is equal to the atomic, molecular, or formula mass of a substance taken in grams.

4.26 (a) $4.51 \text{ g Kr} \left(\dfrac{1 \text{ mol Kr}}{83.8 \text{ g Kr}} \right) = 0.0538 \text{ mol Kr}$

(b) $48.0 \text{ g Mg} \left(\dfrac{1 \text{ mol Mg}}{24.3 \text{ g Mg}} \right) = 1.98 \text{ mol Mg}$

(c) $0.273 \text{ g Ag} \left(\dfrac{1 \text{ mol Ag}}{107.9 \text{ g Ag}} \right) = 0.00253 \text{ mol Ag}$

4.27 (a) $1.37 \text{ mol Fe} \left(\dfrac{55.8 \text{ g Fe}}{1 \text{ mol Fe}} \right) = 76.4 \text{ g Fe}$

(b) $0.783 \text{ mol Si} \left(\dfrac{28.1 \text{ g Si}}{1 \text{ mol Si}} \right) = 22.0 \text{ g Si}$

(c) $0.0915 \text{ mol Ar} \left(\dfrac{39.9 \text{ g Ar}}{1 \text{ mol Ar}} \right) = 3.65 \text{ g Ar}$

4.28 (a) $0.853 \text{ mol Pb} \left(\dfrac{6.02 \times 10^{23} \text{ atoms Pb}}{1 \text{ mol Pb}} \right) = 5.14 \times 10^{23} \text{ atoms Pb}$

(b) $0.637 \text{ mol Ni} \left(\dfrac{6.02 \times 10^{23} \text{ atoms Ni}}{1 \text{ mol Ni}} \right) = 3.83 \times 10^{23} \text{ atoms Ni}$

(c) $3.11 \text{ g Ne} \left(\dfrac{1 \text{ mol Ne}}{20.2 \text{ g Ne}} \right) \left(\dfrac{6.02 \times 10^{23} \text{ atoms Ne}}{1 \text{ mol Ne}} \right) = 9.27 \times 10^{22} \text{ atoms Ne}$

(d) $0.274 \text{ g Ag} \left(\dfrac{1 \text{ mol Ag}}{107.9 \text{ g Ag}} \right) \left(\dfrac{6.02 \times 10^{23} \text{ atoms Ag}}{1 \text{ mol Ag}} \right) = 1.53 \times 10^{21} \text{ atoms Ag}$

4.29 (a) $6.73 \text{ g NaNO}_3 \left(\dfrac{1 \text{ mol NaNO}_3}{85.0 \text{ g NaNO}_3} \right) = 0.0792 \text{ mol NaNO}_3$

(b) $11.4 \text{ g C}_2\text{H}_6\text{O} \left(\dfrac{1 \text{ mol C}_2\text{H}_6\text{O}}{46.0 \text{ g C}_2\text{H}_6\text{O}} \right) = 0.248 \text{ mol C}_2\text{H}_6\text{O}$

(c) $0.855 \text{ g CuSO}_4 \left(\dfrac{1 \text{ mol CuSO}_4}{159.6 \text{ g CuSO}_4} \right) = 0.00536 \text{ mol CuSO}_4$

(d) $0.218 \text{ g F}_2 \left(\dfrac{1 \text{ mol F}_2}{38.0 \text{ g F}_2} \right) = 0.00574 \text{ mol F}_2$

4.30 (a) $0.263 \text{ mol Cl}_2 \left(\dfrac{71.0 \text{ g Cl}_2}{1 \text{ mol Cl}_2} \right) = 18.7 \text{ g Cl}_2$

(b) $3.45 \text{ mol CHCl}_3 \left(\dfrac{119.5 \text{ g CHCl}_3}{1 \text{ mol CHCl}_3} \right) = 412 \text{ g CHCl}_3$

(c) $0.513 \text{ mol AgCl} \left(\dfrac{143.4 \text{ g AgCl}}{1 \text{ mol AgCl}} \right) = 73.6 \text{ g AgCl}$

4.31 (continued)

(d) $0.0125 \text{ mol Na}_2\text{S}_2\text{O}_3 \left(\dfrac{158.2 \text{ g Na}_2\text{S}_2\text{O}_3}{1 \text{ mol Na}_2\text{S}_2\text{O}_3} \right) = 1.98 \text{ g Na}_2\text{S}_2\text{O}_3$

4.31 (a) $0.713 \text{ mol C}_4\text{H}_{10} \left(\dfrac{6.02 \times 10^{23} \text{ molecules C}_4\text{H}_{10}}{1 \text{ mol C}_4\text{H}_{10}} \right) = 4.29 \times 10^{23} \text{ molecules C}_4\text{H}_{10}$

(b) $0.473 \text{ mol O}_2 \left(\dfrac{6.02 \times 10^{23} \text{ molecules O}_2}{1 \text{ mol O}_2} \right) = 2.85 \times 10^{23} \text{ molecules O}_2$

(c) $2.67 \text{ g C}_2\text{H}_6 \left(\dfrac{1 \text{ mol C}_2\text{H}_6}{30.0 \text{ g C}_2\text{H}_6} \right) \left(\dfrac{6.02 \times 10^{23} \text{ molecules C}_2\text{H}_6}{1 \text{ mol C}_2\text{H}_6} \right)$
   $= 5.36 \times 10^{22} \text{ molecules C}_2\text{H}_6$

(d) $1.08 \text{ g C}_6\text{H}_{12}\text{O}_6 \left(\dfrac{1 \text{ mol C}_6\text{H}_{12}\text{O}_6}{180.0 \text{ g C}_6\text{H}_{12}\text{O}_6} \right) \left(\dfrac{6.02 \times 10^{23} \text{ molecules C}_6\text{H}_{12}\text{O}_6}{1 \text{ mol C}_6\text{H}_{12}\text{O}_6} \right)$
   $= 3.61 \times 10^{21} \text{ molecules C}_6\text{H}_{12}\text{O}_6$

4.32   Hydrogen, H_2; nitrogen, N_2; oxygen, O_2; fluorine, F_2; chlorine, Cl_2; bromine, Br_2; and iodine, I_2.

4.33 (a) $3.58 \text{ g Br}_2 \left(\dfrac{1 \text{ mol Br}_2}{159.8 \text{ g Br}_2} \right) = 0.0224 \text{ mol Br}_2$

(b) $0.0224 \text{ mol Br}_2 \left(\dfrac{6.02 \times 10^{23} \text{ molecules Br}_2}{1 \text{ mol Br}_2} \right) = 1.35 \times 10^{22} \text{ molecules Br}_2$

(c) $3.58 \text{ g Br} \left(\dfrac{1 \text{ mol Br}}{79.9 \text{ g Br}} \right) = 0.0448 \text{ mol Br}$

(d) $0.0448 \text{ mol Br} \left(\dfrac{6.02 \times 10^{23} \text{ atoms Br}}{1 \text{ mol Br}} \right) = 2.70 \times 10^{22} \text{ atoms Br}$

4.34 $37.2 \text{ g C} \left(\dfrac{1 \text{ mol C}}{12.0 \text{ g C}} \right) = 3.10 \text{ mol C} \xrightarrow{\div 1.55} 2$

   $7.8 \text{ g H} \left(\dfrac{1 \text{ mol H}}{1.0 \text{ g H}} \right) = 7.8 \text{ mol H} \xrightarrow{\div 1.55} 5$

   $55.0 \text{ g Cl} \left(\dfrac{1 \text{ mol Cl}}{35.5 \text{ g Cl}} \right) = 1.55 \text{ mol Cl} \xrightarrow{\div 1.55} 1$

empirical formula: C_2H_5Cl

4.35 $68.3 \text{ g Pb} \left(\dfrac{1 \text{ mol Pb}}{207.2 \text{ g Pb}} \right) = 0.330 \text{ mol Pb} \xrightarrow{\div\ 0.330} 1$

$10.6 \text{ g S} \left(\dfrac{1 \text{ mol S}}{32.1 \text{ g S}} \right) = 0.330 \text{ mol S} \xrightarrow{\div\ 0.330} 1$

$21.1 \text{ g O} \left(\dfrac{1 \text{ mol O}}{16.0 \text{ g O}} \right) = 1.32 \text{ mol O} \xrightarrow{\div\ 0.330} 4$

empirical formula: $PbSO_4$

4.36 $45.0 \text{ g C} \left(\dfrac{1 \text{ mol C}}{12.0 \text{ g C}} \right) = 3.75 \text{ mol C} \xrightarrow{\div\ 2.50} 1.5 \xrightarrow{\times\ 2} 3$

$7.5 \text{ g H} \left(\dfrac{1 \text{ mol H}}{1.0 \text{ g H}} \right) = 7.5 \text{ mol H} \xrightarrow{\div\ 2.50} 3 \xrightarrow{\times\ 2} 6$

$47.5 \text{ g F} \left(\dfrac{1 \text{ mol F}}{19.0 \text{ g F}} \right) = 2.50 \text{ mol F} \xrightarrow{\div\ 2.50} 1 \xrightarrow{\times\ 2} 2$

empirical formula: $C_3H_6F_2$

4.37 (a) $69.94 \text{ g Fe} \left(\dfrac{1 \text{ mol Fe}}{55.8 \text{ g Fe}} \right) = 1.25 \text{ mol Fe} \xrightarrow{\div\ 1.25} 1 \xrightarrow{\times\ 2} 2$

$30.06 \text{ g O} \left(\dfrac{1 \text{ mol O}}{16.0 \text{ g O}} \right) = 1.88 \text{ mol O} \xrightarrow{\div\ 1.25} 1.5 \xrightarrow{\times\ 2} 3$

empirical formula: Fe_2O_3

(b) $72.36 \text{ g Fe} \left(\dfrac{1 \text{ mol Fe}}{55.8 \text{ g Fe}} \right) = 1.30 \text{ mol Fe} \xrightarrow{\div\ 1.30} 1 \xrightarrow{\times\ 3} 3$

$27.64 \text{ g O} \left(\dfrac{1 \text{ mol O}}{16.0 \text{ g O}} \right) = 1.73 \text{ mol O} \xrightarrow{\div\ 1.30} 1.33 \xrightarrow{\times\ 3} 4$

empirical formula: Fe_3O_4

(c) $77.73 \text{ g Fe} \left(\dfrac{1 \text{ mol Fe}}{55.8 \text{ g Fe}} \right) = 1.39 \text{ mol Fe} \xrightarrow{\div\ 1.39} 1$

$22.27 \text{ g O} \left(\dfrac{1 \text{ mol O}}{16.0 \text{ g O}} \right) = 1.39 \text{ mol O} \xrightarrow{\div\ 1.39} 1$

empirical formula: FeO

4.38 $0.122 \text{ g C}\left(\dfrac{1 \text{ mol C}}{12.0 \text{ g C}}\right) = 0.0102 \text{ mol C} \xrightarrow{\;\div\; 0.0102\;} 1$

$0.578 \text{ g F}\left(\dfrac{1 \text{ mol F}}{19.0 \text{ g F}}\right) = 0.0304 \text{ mol F} \xrightarrow{\;\div\; 0.0102\;} 3$

empirical formula: CF_3

4.39 $0.227 \text{ g Ni}\left(\dfrac{1 \text{ mol Ni}}{58.7 \text{ g Ni}}\right) = 0.00387 \text{ mol Ni} \xrightarrow{\;\div\; 0.00387\;} 1$

$0.619 \text{ g Br}\left(\dfrac{1 \text{ mol Br}}{79.9 \text{ g Br}}\right) = 0.00775 \text{ mol Br} \xrightarrow{\;\div\; 0.00387\;} 2$

empirical formula: $NiBr_2$

4.40 $3.30 \text{ g Cr}\left(\dfrac{1 \text{ mol Cr}}{52.0 \text{ g Cr}}\right) = 0.0635 \text{ mol Cr} \xrightarrow{\;\div\; 0.0635\;} 1 \xrightarrow{\;\times\; 2\;} 2$

$1.52 \text{ g O}\left(\dfrac{1 \text{ mol O}}{16.0 \text{ g O}}\right) = 0.0950 \text{ mol O} \xrightarrow{\;\div\; 0.0635\;} 1.5 \xrightarrow{\;\times\; 2\;} 3$

empirical formula: Cr_2O_3

4.41 molar mass = $\dfrac{12.6 \text{ g}}{0.137 \text{ mol}}$ = 92.0 g/mol

4.42 molar mass = $\dfrac{203 \text{ g}}{2.05 \text{ mol}}$ = 99.0 g/mol

4.43 molar mass = $\dfrac{72.5 \text{ g}}{0.212 \text{ mol}}$ = 342 g/mol

4.44 62.5 u/molecule

4.45 Empirical formula mass = $(1 \times 14.0) + (2 \times 1.0) = 16.0$

$\dfrac{32.0}{16.0} = 2$ molecular formula: N_2H_4

4.46 Empirical formula mass = $(2 \times 12.0) + 1.0 + 19.0 = 44.0$

$\dfrac{132.0}{44.0} = 3$ molecular formula: $C_6H_3F_3$

4.47 (a) $80.0 \text{ g C} \left(\dfrac{1 \text{ mol C}}{12.0 \text{ g C}} \right) = 6.67 \text{ mol C} \xrightarrow{\;\div\, 6.67\;} 1$

$20.0 \text{ g H} \left(\dfrac{1 \text{ mol H}}{1.0 \text{ g H}} \right) = 20 \text{ mol H} \xrightarrow{\;\div\, 6.67\;} 3$

empirical formula: CH_3

(b) molar mass $= \dfrac{3.75 \text{ g}}{0.125 \text{ mol}} = 30.0 \text{ g/mol}$

(c) empirical formula mass $= 12.0 + (3 \times 1.0) = 15.0$

$\dfrac{30.0}{15.0} = 2 \qquad$ molecular formula: C_2H_6

4.48 (a) $49.0 \text{ g C} \left(\dfrac{1 \text{ mol C}}{12.0 \text{ g C}} \right) = 4.08 \text{ mol C} \xrightarrow{\;\div\, 1.36\;} 3$

$2.7 \text{ g H} \left(\dfrac{1 \text{ mol H}}{1.0 \text{ g H}} \right) = 2.7 \text{ mol H} \xrightarrow{\;\div\, 1.36\;} 2$

$48.3 \text{ g Cl} \left(\dfrac{1 \text{ mol Cl}}{35.5 \text{ g Cl}} \right) = 1.36 \text{ mol Cl} \xrightarrow{\;\div\, 1.36\;} 1$

empirical formula: C_3H_2Cl

(b) molar mass $= \dfrac{10.0 \text{ g}}{0.0680 \text{ mol}} = 147 \text{ g/mol}$

(c) empirical formula mass $= (3 \times 12.0) + (2 \times 1.0) + 35.5 = 73.5$

$\dfrac{147}{73.5} = 2 \qquad$ molecular formula: $C_6H_4Cl_2$

4.49 $355 \text{ mg K}_2\text{CrO}_4 \left(\dfrac{1 \text{ g K}_2\text{CrO}_4}{1000 \text{ mg K}_2\text{CrO}_4} \right) \left(\dfrac{1 \text{ mol K}_2\text{CrO}_4}{194.2 \text{ g K}_2\text{CrO}_4} \right) \left(\dfrac{4 \text{ mol O atoms}}{1 \text{ mol K}_2\text{CrO}_4} \right)$

$\times \left(\dfrac{6.02 \times 10^{23} \text{ atoms O}}{1 \text{ mol O atoms}} \right) = 4.40 \times 10^{21} \text{ atoms O}$

4.50 $3.45 \text{ mL Hg} \left(\dfrac{13.6 \text{ g Hg}}{1 \text{ mL Hg}} \right) \left(\dfrac{1 \text{ mol Hg}}{200.6 \text{ g Hg}} \right) \left(\dfrac{6.02 \times 10^{23} \text{ atoms Hg}}{1 \text{ mol Hg}} \right)$

$= 1.41 \times 10^{23} \text{ atoms Hg}$

4.51 (a) $4.13 \text{ g C} \left(\dfrac{1 \text{ mol C}}{12.0 \text{ g C}} \right) = 0.344 \text{ mol C} \xrightarrow{\;\div\, 0.344\;} 1 \xrightarrow{\;\times\, 4\;} 4$

4.51 (continued)

$$0.43 \text{ g H}\left(\frac{1 \text{ mol H}}{1.0 \text{ g H}}\right) = 0.43 \text{ mol H} \xrightarrow{\div 0.344} 1.25 \xrightarrow{\times 4} 5$$

empirical formula: C_4H_5

(b) molar mass $= \dfrac{4.56 \text{ g}}{0.0430 \text{ mol}} = 106 \text{ g/mol}$

(c) empirical formula mass $= (4 \times 12.0) + (5 \times 1.0) = 53.0$

$\dfrac{106}{53.0} = 2$ molecular formula: C_8H_{10}

4.52 (a) $\dfrac{4.00 \text{ g}}{6.02 \times 10^{23} \text{ atoms}} = 6.64 \times 10^{-24} \text{ g/atom}$

(b) $\dfrac{6.64 \times 10^{-24} \text{ g / atom}}{4.00 \text{ u / atom}} = 1.66 \times 10^{-24} \text{ g/u}$

4.53 (a)

$2 \times$ H	=	$2 \times$ 1.0	=	2.0
$1 \times$ O	=	$1 \times$ 16.0	=	16.0
		H_2O	=	18.0

$1 \times$ Rh	=	$1 \times$ 102.9	=	102.9
$3 \times$ Cl	=	$3 \times$ 35.5	=	106.5
		$RhCl_3$	=	209.4

$RhCl_3 \cdot 6 \text{ } H_2O = 209.4 + 6(18.0) = 317.4$

One gram of $RhCl_3 \cdot 6 \text{ } H_2O$ would cost \$8.00. The value of the rhodium in it would be:

$$1.00 \text{ g } RhCl_3 \cdot 6 \text{ } H_2O\left(\frac{102.9 \text{ g Rh}}{317.4 \text{ g } RhCl_3 \cdot 6 \text{ } H_2O}\right)\left(\frac{\$20.00}{1 \text{ g Rh}}\right) = \$6.48$$

It would not be wise to spend \$8.00 on a sample of a rhodium compound that is only worth \$6.48.

4.54 First find the molar mass of each compound:

$1 \times$ C	=	$1 \times$ 12.0	=	12.0
$2 \times$ H	=	$2 \times$ 1.0	=	2.0
$2 \times$ Cl	=	$2 \times$ 35.5	=	71.0
		CH_2Cl_2	=	85.0

$8 \times$ C	=	$8 \times$ 12.0	=	96.0
$10 \times$ H	=	$10 \times$ 1.0	=	10.0
$4 \times$ N	=	$4 \times$ 14.0	=	56.0
$2 \times$ O	=	$2 \times$ 16.0	=	32.0
		$C_8H_{10}N_4O_2$	=	194.0

For CH_2Cl_2: 12.0 g C = 85.0 g CH_2Cl_2
For $C_8H_{10}N_4O_2$: 96.0 g C = 194.0 g $C_8H_{10}N_4O_2$

$$1.00 \text{ g } CH_2Cl_2\left(\frac{12.0 \text{ g C}}{85.0 \text{ g } CH_2Cl_2}\right)\left(\frac{194.0 \text{ g } C_8H_{10}N_4O_2}{96.0 \text{ g C}}\right) = 0.285 \text{ g } C_8H_{10}N_4O_2$$

4.55 $1 \times$ Na $= 1 \times 23.0 = 23.0$
$\underline{1 \times \text{Cl} = 1 \times 35.5 = 35.5}$
NaCl $= 58.5$

$$1.00 \text{ T} \left(\frac{3 \text{ tsp}}{1 \text{ T}} \right) \left(\frac{7.8 \text{ g NaCl}}{1 \text{ tsp}} \right) \left(\frac{23.0 \text{ g Na}}{58.5 \text{ g NaCl}} \right) \left(\frac{1000 \text{ mg Na}}{1 \text{ g Na}} \right)$$

$= 9200$ mg Na $(9.2 \times 10^3$ mg Na) for the entire meatloaf

$$\frac{9200 \text{ mg Na}}{5 \text{ persons}} = 1800 \text{ mg/person} \ (1.8 \times 10^3 \text{ mg/person})$$

$$\frac{1800 \text{ mg}}{1100 \text{ mg}} \times 100\% = 160\% \text{ of minimum RDA guideline}$$

$$\frac{1800 \text{ mg}}{3300 \text{ mg}} \times 100\% = 55\% \text{ of maximum RDA guideline}$$

4.56 Whenever n is an even number, the molecular formula will be divisible by 2, and will therefore not be the empirical formula.

4.57 Since water does not contain any carbon, when glucose and fructose combine (giving off water), the resulting sucrose molecule must have a higher percentage of carbon than either of the smaller sugars that combined.

4.58 The percentage of fluorine (or of xenon) will differ in each of the three compounds. We can calculate the percentage of these elements in each of the three compounds using the method illustrated in this chapter. If we are given a sample of one of them and wish to know which it is, we may simply have it analyzed for its percentage composition and match the result with one of our three calculations.

CHAPTER 5

Atomic Structure

5.1 (a) 4 u (b) +2

5.2 (a) 10 (b) 29 (c) 83 (d) 7 (e) 51 (f) 80

5.3 (a) Beryllium-9 ^9_4Be (b) Carbon-12 $^{12}_6C$

  (c) Oxygen-18 $^{18}_8O$ (d) Chlorine-35 $^{35}_{17}Cl$

  (e) Uranium-238 $^{238}_{92}U$ (f) Lawrencium-257 $^{257}_{103}Lr$

  (g) Carbon-13 $^{13}_6C$ (h) Chlorine-37 $^{37}_{17}Cl$

5.4 Carbon-12 and carbon-13 are isotopes of one another, as are chlorine-35 and chlorine-37.

5.5

Name	Symbol	Atomic Number	Number of Protons	Number of Electrons	Number of Neutrons	Mass Number
Lithium	Li	3	3	3	4	7
Carbon	C	6	6	6	7	13
Nitrogen	N	7	7	7	8	15
Gold	Au	79	79	79	118	197
Lead	Pb	82	82	82	124	206
Uranium	U	92	92	92	143	235
Uranium	U	92	92	92	146	238
Neptunium	Np	93	93	93	144	237
Plutonium	Pu	94	94	94	148	242

5.6 (a) 3:1 (b) 75,770 (c) 40 (d) 1; nitrogen-14; yes
  (e) Beryllium, fluorine, sodium, aluminum, and phosphorus each has only one naturally occurring isotope. For each of these elements, the atomic mass is approximately equal to the mass number of that naturally occurring isotope.
  (f) Potassium-40 has 19 protons and 21 neutrons, whereas calcium-40 has 20 protons and 20 neutrons.
  (g) Hydrogen-2 and helium-3 (1 each); beryllium-9 and boron-10 (5 each); carbon-13 and nitrogen-14 (7 each); oxygen-18, fluorine-19, and neon-20 (10 each); silicon-30 and phosphorus-31 (16 each); potassium-39 and calcium-40 (20 each)

5.7 Atomic mass $= (0.6917)(62.9296\ u) + (0.3083)(64.9278\ u)$
    $= 43.53\ u + 20.02\ u = 63.55\ u$

5.8 Atomic mass $= (0.5184)(106.9051\ u) + (0.4816)(108.9048\ u)$
    $= 55.42\ u + 52.45\ u = 107.87\ u$

5.9 (a) 47 protons, 46 electrons   (b) 26 protons, 23 electrons
(c) 30 protons, 28 electrons   (d) 50 protons, 46 electrons
(e) 34 protons, 36 electrons   (f) 15 protons, 18 electrons

5.10 (a) Hg^{2+} (b) Cu^+ (c) Rb^+ (d) Ni^{2+} (e) Au^{3+} (f) Te^{2-}

5.11 *Ideas that are still accepted:* All matter is composed of atoms. The atoms of each element are different from atoms of all other elements.
Ideas that had to be modified: Atoms are not indivisible; they are composed of protons, neutrons, and electrons. Atoms of the same element may differ in their masses.

5.12 Rutherford demonstrated that an atom is composed of a very dense, positively-charged nucleus, surrounded by a largely empty volume in which the electrons are found. The radius of the nucleus is about 1/100,000 that of the atom. Virtually the entire mass of an atom is concentrated in the nucleus.

5.13 proton: +1, 1 u; neutron: 0, 1 u; electron: –1, 0 u

5.14 (a) 12 u; (b) 14 u; (c) 14 u; (d) atoms (a) and (b).

5.15 (a) The *atomic number* is the number of protons in the nucleus of an atom.
(b) The *mass number* is the sum of the protons and neutrons in the nucleus of an atom.
(c) An *isotope* is one of two or more atoms of the same element that differ in mass number.

5.16 (a) lithium-7 $^{7}_{3}Li$ (b) carbon-14 $^{14}_{6}C$
(c) fluorine-19 $^{19}_{9}F$ (d) sulfur-33 $^{33}_{16}S$
(e) iron-56 $^{56}_{26}Fe$ (f) plutonium-238 $^{238}_{94}Pu$
(g) iron-58 $^{58}_{26}Fe$ (h) sulfur-31 $^{31}_{16}S$
(i) The pairs of isotopes are: (d) & (h), and (e) & (g).

5.17

Name	Symbol	Atomic Number	Number of Protons	Number of Electrons	Number of Neutrons	Mass Number
boron	B	5	5	5	6	11
oxygen	O	8	8	8	9	17
chlorine	Cl	17	17	17	18	35
phosphorus	P	15	15	15	16	31
calcium	Ca	20	20	20	28	48
tin	Sn	50	50	50	69	119
polonium	Po	84	84	84	126	210
radon	Rn	86	86	86	136	222
nobelium	No	102	102	102	152	254

5.18 (b)

5.19 (a)

5.20 (a)

5.21 atomic mass   = (0.60108)(68.9256 u) + (0.39892)(70.9247 u)
                       = 41.430 u + 28.293 u = 69.723 u

5.22 atomic mass   = (0.72165)(84.9118 u) + (0.27835)(86.9092 u)
                       = 61.277 u + 24.191 u = 85.468 u

5.23 (a) $37p^+, 36e^-$ (b) $27p^+, 24e^-$ (c) $80p^+, 78e^-$ (d) $82p^+, 78e^-$ (e) $52p^+, 54e^-$
(f) $7p^+, 10e^-$

5.24 (a) Zn^{2+} (b) Cr^{3+} (c) Ga^{3+} (d) Cs^+ (e) Pt^{4+} (f) Sb^{3-}

5.25 An *ion* is an atom or a group of atoms bearing a net charge other than zero. A monatomic ion has only one atom; a polyatomic ion has two or more atoms.

5.26 The ratio of atomic size : nuclear size is 10^{-10} m : 10^{-14} m or 10,000 : 1.
If the nucleus in the model is 2 cm, then the atom as a whole must be 20,000 cm.

$$20,000 \text{ cm} \left(\frac{1 \text{ in.}}{2.54 \text{ cm}} \right) \left(\frac{1 \text{ ft}}{12 \text{ in.}} \right) \left(\frac{1 \text{ yd}}{3 \text{ ft}} \right) = 200 \text{ yd}$$

A football or baseball stadium is roughly 200 yards in diameter. Thus, a domed stadium might be a good model for half of an atom whose nucleus is the size of a marble.

5.27 17 neutrons (15 protons and 17 neutrons give a mass number of 32)

5.28 Since bromine is roughly a 50:50 mixture of the isotopes, a group of 4 molecules would have the following statistical distribution of masses:

carbon	+	two hydrogens	+	1st bromine	+	2nd bromine	=	total mass
12 u	+	2(1 u)	+	79 u	+	79 u	=	172 u
12 u	+	2(1 u)	+	79 u	+	81 u	=	174 u
12 u	+	2(1 u)	+	81 u	+	79 u	=	174 u
12 u	+	2(1 u)	+	81 u	+	81 u	=	176 u

Thus, the approximate composition would be 25% with masses of 172 u, 50% with masses of 174 u, and 25% with masses of 176 u.

5.29 140 lb. Normal water, H_2O, has a molecular mass of approximately 18 u: 2(1 u) + 16 u = 18 u. Heavy water, D_2O, has a molecular mass of approximately 20 u: 2(2 u) + 16 u = 20 u. Thus, the mass ratio of heavy water to normal water is 20/18. If 70% of the woman's weight is water, then 91 lb (0.70 × 130 lb) is water and 39 lb is "non-water" matter. Replacement of 91 lb of regular water with heavy water would require 91 lb × (20/18) = 101 lb of heavy water. Adding this to the 39 lb of "non-water" matter gives a mass of 140 lb (101 lb + 39 lb).

5.30 (a) $19p^+$, $18e^-$ (b) $17p^+$, $18e^-$ (c) $16p^+$, $18e^-$

(d) $20p^+$, $18e^-$ (e) all have the same number of electrons.

5.31 26 protons, 30 neutrons, 24 electrons

5.32 (a) 19 protons, 22 neutrons, 18 electrons   (b) 16 protons, 16 neutrons, 18 electrons

(c) 13 protons, 14 neutrons, 10 electrons   (d) 53 protons, 74 neutrons, 54 electrons

5.33 (a) $^{56}_{26}Fe^{3+}$ (b) $^{59}_{27}Co$ (c) $^{56}_{26}Fe^{2+}$ (d) $^{58}_{27}Co^{3+}$ (e) b and d (f) a and c

(g) a and d (h) b

5.34 (a) 1:1   (b) 1:1   (c) 1:3   (d) 3:2

5.35 (a) monatomic ion (b) element (c) compound (d) polyatomic ion

(e) monatomic ion (f) polyatomic ion (g) compound (h) element

CHAPTER 6

Electronic Configuration

6.1 (a) T (b) F (c) T (d) T (e) F (f) T

6.2 (a) F (b) T (c) T (d) F (e) F

6.3 (a) 10 (b) 2 (c) 6 (d) 14 (e) 22 (f) 18 (g) 32

6.4 $3d \rightarrow 4p$; $5s \rightarrow 4d$; $6s \rightarrow 4f$; $4d \rightarrow 5p$; $4f \rightarrow 5d$

6.5
(a) Na: $1s^2 2s^2 2p^6 3s^1$ or   [Ne]$3s^1$
(b) Mg: $1s^2 2s^2 2p^6 3s^2$ or   [Ne]$3s^2$
(c) Al: $1s^2 2s^2 2p^6 3s^2 3p^1$ or [Ne]$3s^2 3p^1$
(d) Si: $1s^2 2s^2 2p^6 3s^2 3p^2$ or [Ne]$3s^2 3p^2$
(e) P: $1s^2 2s^2 2p^6 3s^2 3p^3$ or [Ne]$3s^2 3p^3$
(f) S: $1s^2 2s^2 2p^6 3s^2 3p^4$ or [Ne]$3s^2 3p^4$
(g) Cl: $1s^2 2s^2 2p^6 3s^2 3p^5$ or [Ne]$3s^2 3p^5$
(h) Ar: $1s^2 2s^2 2p^6 3s^2 3p^6$ or [Ne]$3s^2 3p^6$

6.6
(a) P: $1s^2 2s^2 2p^6 3s^2 3p^3$ or [Ne]$3s^2 3p^3$
(b) Ni: $1s^2 2s^2 2p^6 3s^2 3p^6 4s^2 3d^8$ or [Ar]$4s^2 3d^8$
(c) Sr: $1s^2 2s^2 2p^6 3s^2 3p^6 4s^2 3d^{10} 4p^6 5s^2$ or [Kr]$5s^2$
(d) Hg: $1s^2 2s^2 2p^6 3s^2 3p^6 4s^2 3d^{10} 4p^6 5s^2 4d^{10} 5p^6 6s^2 4f^{14} 5d^{10}$ or [Xe]$6s^2 4f^{14} 5d^{10}$
(e) I: $1s^2 2s^2 2p^6 3s^2 3p^6 4s^2 3d^{10} 4p^6 5s^2 4d^{10} 5p^5$ or [Kr]$5s^2 4d^{10} 5p^5$
(f) Eu: $1s^2 2s^2 2p^6 3s^2 3p^6 4s^2 3d^{10} 4p^6 5s^2 4d^{10} 5p^6 6s^2 4f^7$ or [Xe]$6s^2 4f^7$

6.7 filled = 2 electrons; half-filled = 1 electron; empty = 0 electrons.
(a) one filled orbital
(b) two half-filled orbitals, one empty orbital
(c) two filled orbitals, one half-filled orbital
(d) five half-filled orbitals
(e) three filled orbitals, two half-filled orbitals

6.8 (a) 2 (b) 2 (c) 1 (d) 2 (e) 3 (f) 3

6.9 (a)

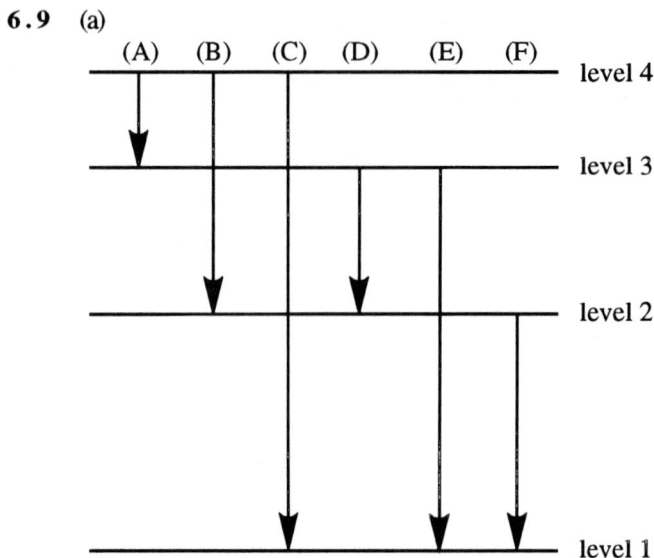

Six different frequencies of light are possible.

(b)

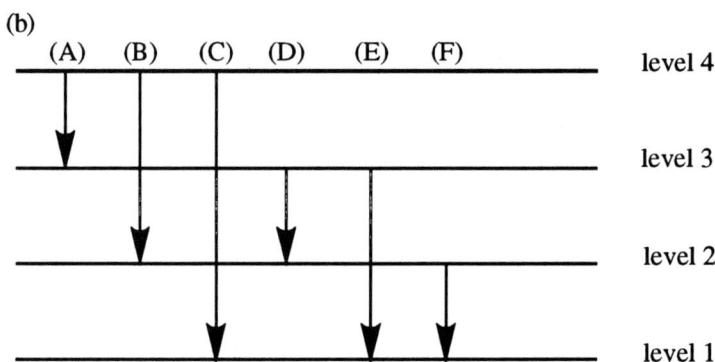

There are only three different frequencies of light. Transitions (D) and (F) are of the same energy as transition (A), and transition (E) is of the same energy as transition (B). Thus transitions (A), (B), and (C) account for the three observed frequencies.

6.10 In the Bohr model, electrons orbit the nucleus in fixed radii. The higher the energy level, the greater the distance from the nucleus. For any energy level, the "average" distance equals the fixed radius. In the quantum-mechanical model, electrons exist in orbitals which describe the probability of finding an electron at given locations, relative to the nucleus. Unlike the Bohr model, the electrons in an orbital are not restricted to fixed distances from the nucleus; but instead they may be found over a wide range of distances. Nevertheless, for higher energy orbitals, there is a greater probability of finding an electron further from the nucleus. Thus, the average electronic distance from the nucleus increases with increasing energy in both the Bohr model and the quantum-mechanical model.

6.11 (a) true   (b) false   (c) true   (d) true   (e) false   (f) true

6.12 (a) five   (b) $4s$, $4p$, $4d$, and $4f$ (c) seven   (d) two   (e) ten   (f) eighteen

6.13 The nine orbitals in a g-sublevel would accomodate 18 electrons.

6.14 Electrons fill atomic orbitals in order of increasing orbital energy.

6.15 You will have to do this on your own. Refer to the legend that accompanies Figure 6-8 for a helpful suggestion.

6.16 Within each principal energy level, the sublevels are in the order $s < p < d < f$. Both the principal energy level and the type of sublevel contribute to the overall energy of a sublevel. The $5s$ sublevel (which has the lowest energy of any of the sublevels in principal energy level 5) is below the $4d$ (which is one of the higher sublevels in energy level 4).

6.17 (a) $1s^2 2s^2 2p^6 3s^1$ or $[Ne]3s^1$
      (b) $1s^2 2s^2 2p^6 3s^2 3p^6 4s^2 3d^{10} 4p^1$ or $[Ar]4s^2 3d^{10} 4p^1$
      (c) $1s^2 2s^2 2p^6 3s^2 3p^6 4s^2$ or $[Ar]4s^2$
      (d) $1s^2 2s^2 2p^6 3s^2 3p^6 4s^2 3d^3$ or $[Ar]4s^2 3d^3$
      (e) $1s^2 2s^2 2p^6 3s^2 3p^6 4s^2 3d^{10} 4p^6 5s^2 4d^{10} 5p^6 6s^2 4f^5$ or $[Xe]6s^2 4f^5$
      (f) $1s^2 2s^2 2p^6 3s^2 3p^6 4s^2 3d^{10} 4p^6 5s^2 4d^2$ or $[Kr]5s^2 4d^2$

6.18 Electrons half-fill the orbitals of a given sublevel before pairing of electrons occurs. Each orbital in the $4f$-sublevel of Eu is half-filled.

6.19 (a) two filled orbitals, one half-filled orbital
      (b) five half-filled orbitals
      (c) five half-filled orbitals, two empty orbitals

6.20 $c = \lambda \nu$: The velocity of light is a constant; frequency and wavelength are inversely related.
      $E = h\nu$: As the frequency increases, so does the energy of each photon.

6.21 Each photon in the radar frequency range has a low energy. However, the cumulative energy for a large number of photons can be large enough to be harmful. The practice of cleaning the radar sending units can be compared to putting one's ear to a loud speaker. Although sound waves of normal intensity are harmless, damage can be done when the intensity (volume) becomes excessive. Similarly, a large number of photons (with their accompanying energy) may be absorbed by the body, leading to illness, when a person stands too close to the source of the radar waves.

6.22 A *continuous spectrum* is like a rainbow. All frequencies are present. A *discontinuous spectrum* has only certain frequencies present. Only certain frequencies of light are present in the hydrogen spectrum because the electrons may only make transitions between the specific energy levels present, giving off light waves of frequencies that correspond to the given energy differences..

6.23 In the Bohr model, electrons circulate around the nucleus in orbits of fixed radius, in a fashion similar to the behavior of planets circulating around the sun. According to this model, an electron can change energy levels by absorbing or releasing an amount of energy corresponding to the energy difference between the initial and final energy states of the transition. An energy change of this type would be accompanied by a change in the radius of the orbit. Most of the Bohr model is correct, especially the concept of electrons existing in discrete energy states and being able to absorb or release only amounts of energy corresponding to the differences in the various energy states. However, electrons do not circulate around the nucleus in orbits of fixed radii. Instead, electronic behavior is described in terms of orbitals for which we may describe the probability of finding the electron at any given location.

6.24 An *orbital* is an energy state that describes the probability of finding an electron at various locations with respect to the nucleus. The location of an electron in an orbital is only described in terms of probabilities. The Bohr model proposed that electrons orbit the nucleus at fixed distances.

6.25 (a) $Li = [He]2s^1$, $Na = [Ne]3s^1$, $K = [Ar]4s^1$
   (b) All end in s^1.
   (c) All are located in Group IA.
   (d) $O = [He]2s^22p^4$, $S = [Ne]3s^23p^4$, $Se = [Ar]4s^23d^{10}4p^4$
        All end in p^4 and are located in group VIA.

6.26 (a) $1s^22s^22p^63s^1$
   (b) $1s^22s^22p^6$
   (c) Ne
   (d) $1s^22s^22p^63s^23p^6$
   (e) Ar
   (f) Both are located in group VIIIA (18).
   (g) Magnesium will lose two electrons to have ten electrons, the same number as neon. Thus, it will form Mg^{2+}.
   (h) Sulfur will gain two electrons to have eighteen electrons, the same number as argon. Thus, it will form S^{2-}.
   (i) Aluminum will lose three electrons to have ten electrons, the same number as neon. Thus, it will form Al^{3+}.

6.27 Each sublevel that follows the filling of a p-sublevel begins a new principal energy level. The filling of each p-sublevel corresponds to the end of a row of the periodic table. The beginning of each new principal energy level corresponds to the beginning of a new row of the periodic table.

6.28 A photon of green light has a higher energy than a photon or red light. Thus, the electronic transitions that correspond to the green light must span a greater energy difference than those for red light.

6.29 The ground-state atom has the fourth principal energy level as the highest occupied level. Since the excited atom has an electron in the fifth principal energy level (the $5s^1$), it must be larger than the ground-state atom.

6.30 If Hund's rule is obeyed, there will be unpaired electrons for all transition metals except those that have a d^{10} configuration. Thus, the only transition metals that sould be incapable of attraction by a magnetic field are zinc, cadmium, and mercury.

6.31 93 million miles = 9.3×10^7 mi;   1 km = 0.621 mi

$$9.3 \times 10^7 \text{ mi} \left(\frac{1 \text{ km}}{0.621 \text{ mi}} \right) \left(\frac{1000 \text{ m}}{1 \text{ km}} \right) \left(\frac{1 \text{ sec}}{3.00 \times 10^8 \text{ m}} \right) = 5.0 \times 10^2 \text{ sec} \quad \text{(or 500 sec)}$$

6.32 (Use "lt-yr" to stand for light-year.)

(a) $1 \text{ lt-yr} = 365 \text{ day} \left(\frac{24 \text{ hr}}{1 \text{ day}} \right) \left(\frac{60 \text{ min}}{1 \text{ hr}} \right) \left(\frac{60 \text{ sec}}{1 \text{ min}} \right) \left(\frac{3.00 \times 10^8 \text{ m}}{1 \text{ sec}} \right) \left(\frac{1 \text{ km}}{1000 \text{ m}} \right)$

$= 9.46 \times 10^{12}$ km

(b) $4.3 \text{ lt-yr} \left(\frac{9.46 \times 10^{12} \text{ km}}{1 \text{ lt - yr}} \right) = 4.1 \times 10^{13}$ km

(c) $9.3 \times 10^7 \text{ mi} \left(\frac{1 \text{ km}}{0.621 \text{ mi}} \right) \left(\frac{1 \text{ lt - yr}}{9.46 \times 10^{12} \text{ km}} \right) = 1.6 \times 10^{-5}$ lt-yr

6.33 The following answer is calculated as of November 22, 1994, 11.0 years after the assassination. To obtain the distance on the day you are solving this problem, add the number of years that have elapsed since that date to the italicized number in the following set-up. (Use "lt-yr" to stand for light-year.)

$$11.0 \text{ lt-yr} \left(\frac{9.46 \times 10^{12} \text{ km}}{1 \text{ lt - yr}} \right) = 1.04 \times 10^{14} \text{ km}$$

CHAPTER 7

The Periodic Table

7.1 (a) The atomic mass of an element represents the average mass of the isotopes that occur naturally. The average mass of the isotopes that make up tellurium is greater than that of the isotopes that make up iodine, even though each tellurium atom has one less proton than each atom of iodine. The greater average mass must mean that the average number of neutrons in a sample of tellurium must be greater than in a sample of iodine.

(b) Argon (atomic number 18) has a greater atomic mass than potassium (atomic number 19). Cobalt (atomic number 27) has a greater atomic mass than nickel (atomic number 28).

7.2 (a) Refer to the glossary.

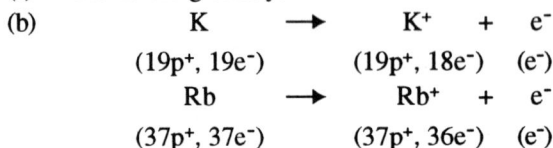

(b)
$$K \longrightarrow K^+ + e^-$$
$$(19p^+, 19e^-) \quad (19p^+, 18e^-) \quad (e^-)$$
$$Rb \longrightarrow Rb^+ + e^-$$
$$(37p^+, 37e^-) \quad (37p^+, 36e^-) \quad (e^-)$$

(c) Refer to the glossary.

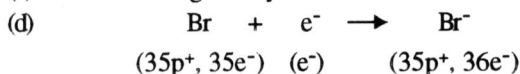

(d)
$$Br + e^- \longrightarrow Br^-$$
$$(35p^+, 35e^-) \quad (e^-) \quad (35p^+, 36e^-)$$

7.3 (a) K^+ (b) F^- (c) Ca^{2+} (d) Se^{2-} (e) no ion (f) Cs^+ (g) O^{2-} (h) no ion (i) Te^{2-}

7.4 (a) alkaline earth metal (b) noble gas (c) halogen (d) alkali metal
(e) chalcogen (f) noble gas (g) halogen (h) alkali metal
(i) alkaline earth metal (j) alkali metal (k) chalcogen

7.5 (a) 2, 3 (b) 1 (c) 2 (d) 5 (e) 1, 3 (f) 1 (g) 6 (h) 6

7.6 Refer to the answer to Problem 6.6.

7.7 (a) Br_2 (b) Te and Po (c) Hg (d) F_2 and Cl_2 (e) B
(f) He (g) I_2

7.8 (a) K (b) Cl (c) S (d) K (e) K

7.9 (a) He and Li^+ (b) F^- and Mg^{2+} (c) Ca^{2+} and P^{3-}

7.10 (a) K^+ (b) Br^- (c) S^{2-} (d) F^- (e) Ar (f) Cl^- (g) Na^+ (h) O^{2-}

7.11 Each of these ions has 18 electrons. However, the nuclear charge increases in the order $S^{2-} \longrightarrow Cl^- \longrightarrow K^+$. This draws the electrons closer to the nucleus, making K^+ the smallest of the three and S^{2-} the largest.

7.12 (a) Elements are arranged in order of atomic number.
     (b) A *period* is a horizontal row of the periodic table.
     (c) A *family* or *group* is a vertical column of the periodic table.

7.13 (a) *Electrostatic forces* are forces that result from the attraction or repulsion of charged particles.
     (b) Like charges repel. Unlike charges attract.
     (c) Electrostatic forces increase with increasing charge.
     (d) Electrostatic forces decrease with increasing distance between the charged bodies.
     (e) *Ionization energy* is the energy required to remove an electron from an atom.
     (f) The negatively-charged electron is attracted to the positively-charged nucleus. Consequently, it takes energy to pull them apart.

7.14 (a) *Electron affinity* is the energy liberated when an electron is captured by a neutral atom.
     (b) $Cl + e^- \longrightarrow Cl^-$
     (c) The electron is attracted to the nucleus. Since energy would be required to separate the electron from the nucleus, energy must be released when it is captured.

7.15 (a) All end in p^6.
     (b) The next element begins filling the next principle energy level.
     (c) The ionization energies of the noble gases are high.
     (d) Loss of an electron leads to a noble gas configuration.
     (e) Gain of an electron leads to a noble gas configuration.

7.16 (a) alkali metal, Li^+ (b) alkaline earth metal, Mg^{2+}
     (c) halogen, Cl^- (d) alkaline earth metal, Ba^{2+}
     (e) chalcogen, S^{2-} (f) halogen, I^-
     (g) alkali metal, Rb^+ (h) alkaline earth metal, Sr^{2+}
     (i) noble gas, no ion   (j) halogen, F^-
     (k) alkali metal, Na^+ (l) halogen, Br^-
     (m) noble gas, no ion   (n) chalcogen, Po^{2-}

7.17 (a) 1 (b) 4 (c) 2 (d) 6 (e) 1, 3, 6 (f) 2, 6 (g) 4 (h) 3,6 (i) 3, 5, 6 (j) 5

7.18 Transition metals: a, b, d, f, i, j. Rare earth metals: c, e, g, h.

7.19 (a) they fill d sublevels   (b) they fill f sublevels.

7.20 See answer to problem 6.17.

7.21 (a) shiny, malleable, conduct electricity
     (b) metal: Al, Hg; semimetal: Ge, As; nonmetal: N_2, I_2.

7.22 (a) Atomic size increases going down a column.
     (b) The principal energy level of the highest filled orbital increases, and hence the size of the atom increases as well.

7.23 (a) Atomic size decreases going across each period from left to right.
  (b) The effective charge equals the nuclear charge minus the electrons in the last noble gas core.
  (c) The effective charge increases going from left to right across each row. This draws the outer electrons closer to the nucleus going across each row.

7.24 (a) The size of a cation is smaller than its parent atom.
  (b) The size of an anion is larger than its parent atom.
  (c) *Isoelectronic* species have the same electronic configurations.
  (d) The sizes of isoelectronic species decrease with increasing nuclear charge (atomic number); the same number of electrons are drawn closer to the more positively-charged nucleus.

7.25 (a) $O < Se < Po$ (b) $Fe < Ru < Os$ (c) $Cl < Al < Mg$
  (d) $Br < Ca < K$ (e) $Cl < Sn < Cs$ (f) $Br < As < Sb$
  (g) $K^+ < Cl^- < P^{3-}$ (h) $Mg^{2+} < Mg < Na$ (i) $F < F^- < O^{2-}$

7.26 (a) Al^{3+} (b) P^{3-}

7.27 (a) 2 to form Mg^{2+}; (b) 2 to form S^{2-}; (c) 1:1; (d) 1 to form Na^+, 2 to form O^{2-}, 2:1;
  (e) 3 to form Al^{3+}, 2 to form S^{2-}, 2:3.

7.28 (a) $NaCl$ (b) $MgBr_2$ (c) AlI_3 (d) Al_2O_3

7.29 (a) As the effective charge increases (going across a given period), the electrons are drawn more closely to the nucleus. Thus, yttrium (with a smaller effective charge) is larger than zirconium (with a larger effective charge).
  (b) In a similar fashion, scandium (atomic number 21) is expected to be larger than titanium (atomic number 22).

7.30 The effective charge on all alkali metals is +1. The effective charge increases going across each row. Since sodium is larger than any noble gas, it means that as the effective charge increases, the size of the atom shrinks significantly. Even the largest noble gas is smaller than sodium, an element with a relatively low atomic number (11). Thus, effective charge is an extremely important factor in determining atomic size.

7.31 The property of malleability is the ability to be pounded and bent into different shapes. Materials that are used to fill dental cavities must be capable of being shaped easily to fill all of the crevices of the cavity as well as to be carved to simulate the natural shape of the tooth. By increasing the malleability of silver, mercury improves its qualities as a material for this purpose.

7.32 When lithium loses its first electron, it forms a noble gas configuration. Thus, loss of a second electron is extremely difficult, and the second ionization energy is very large. However, berylium still has one electron beyond the last noble gas core after losing its first electron. Thus, it loses its second electron readily and has a correspondingly low second ionization energy.

CHAPTER 8

Chemical Bonding

8.1 Elements in group IA have one valence electron, those in group IIA have two valence electrons, and so forth. Those elements in group VIIIA (the noble gases) have eight valence electrons (with the exception of helium, which has two).

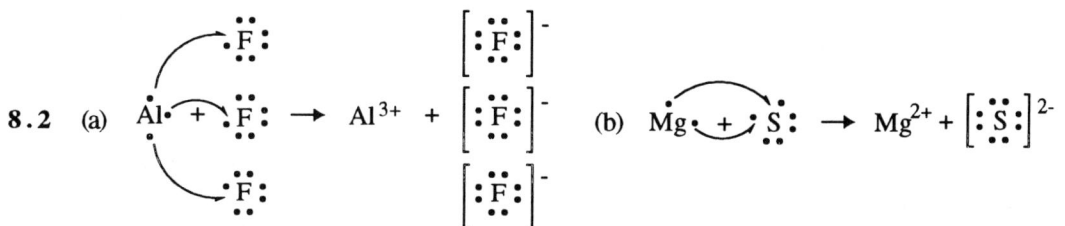

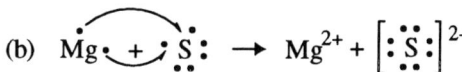

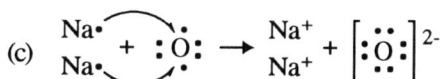

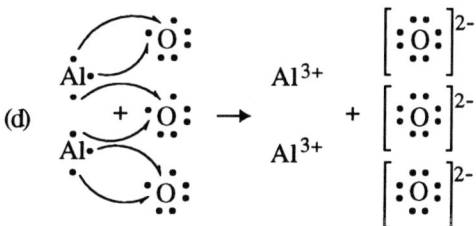

8.2 (a) Al· + ·F̈: → Al³⁺ + [:F̈:]⁻ ... [:F̈:]⁻ ... [:F̈:]⁻

(b) Mg· + ·S̈: → Mg²⁺ + [:S̈:]²⁻

(c) Na· + :Ö: → Na⁺ + [:Ö:]²⁻
Na·   Na⁺

(d) Al· + ·Ö: → Al³⁺ + [:Ö:]²⁻
Al·   Al³⁺   [:Ö:]²⁻
                   [:Ö:]²⁻

8.3 (a) covalent (b) ionic (c) covalent (d) covalent (e) ionic (f) covalent

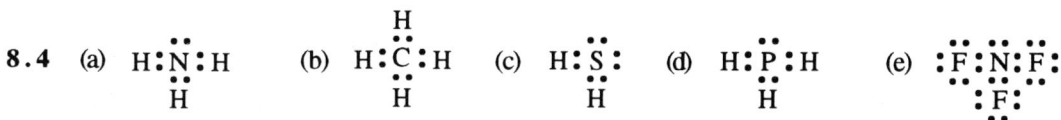

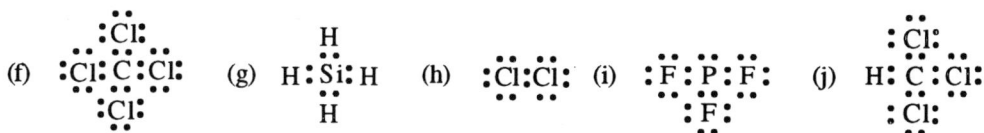

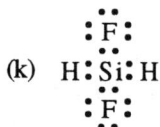

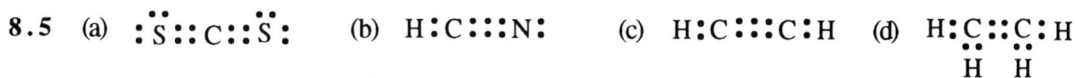

8.4 (a) H:N̈:H with H below
(b) H:C:H with H above and H below
(c) H:S̈: with H below
(d) H:P̈:H with H below
(e) :F̈:N̈:F̈: with :F̈: below

(f) :C̈l:C:C̈l: with :C̈l: above and :C̈l: below
(g) H:S̈i:H with H above and H below
(h) :C̈l:C̈l:
(i) :F̈:P̈:F̈: with :F̈: below
(j) H:C:C̈l: with :C̈l: above and :C̈l: below

(k) :F̈: above, H:S̈i:H, :F̈: below

8.5 (a) :S̈::C::S̈:
(b) H:C:::N:
(c) H:C:::C:H
(d) H:C::C:H with H H below

(e) :Ö: above, H:C:H

8.6 (a)  (b) 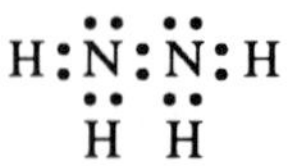 (c)  (d) 

(e) 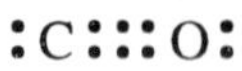 (f)  (g) 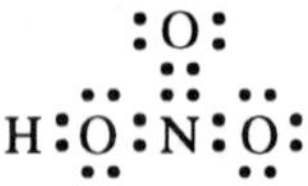 (h) 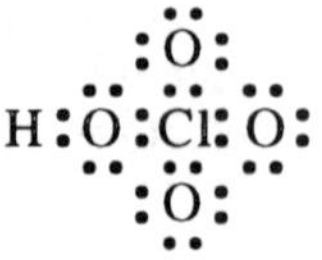

8.7 From problem 8.4:

(a) 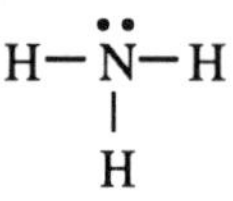 (b) 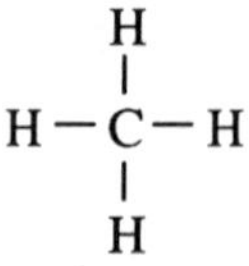 (c) 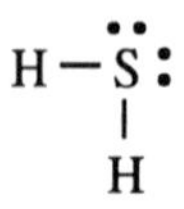 (d) 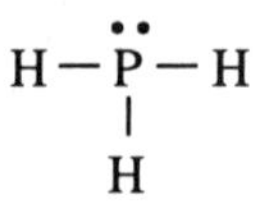

(e) 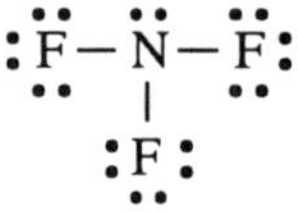 (f) 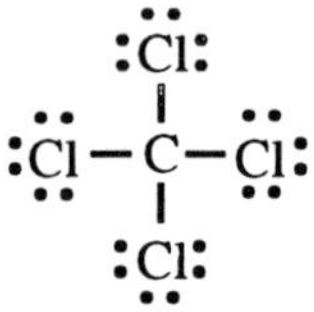 (g) 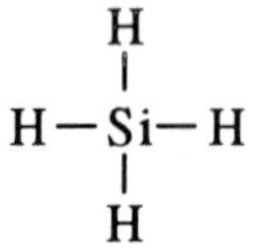 (h) 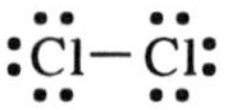

(i) 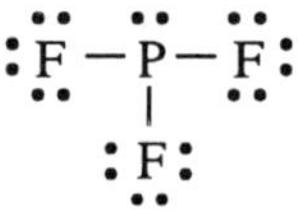 (j) 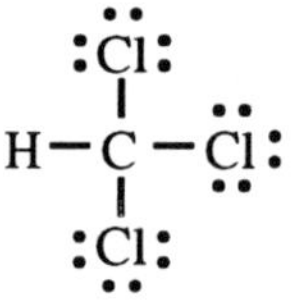 (k) 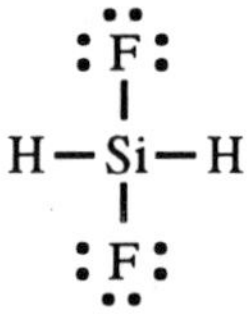

From problem 8.5:

(a) 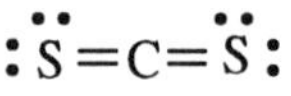 (b) 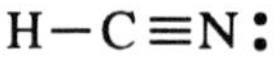 c) 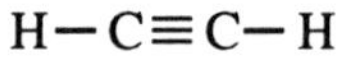 (d) 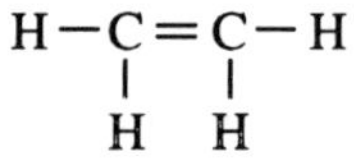

(e) 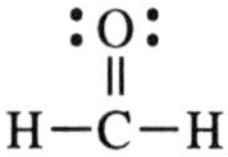

8.8 (a) 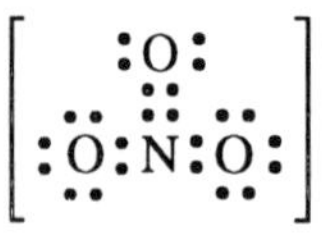 (b)  (c) 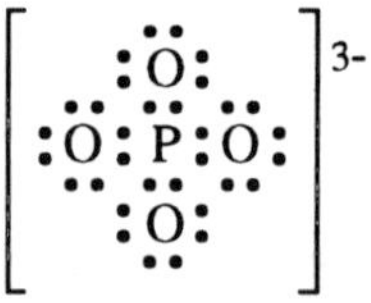 (d) 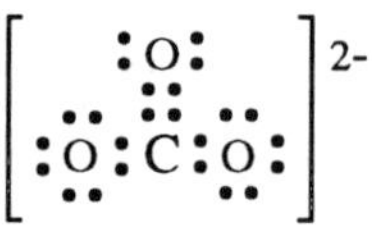

(e) 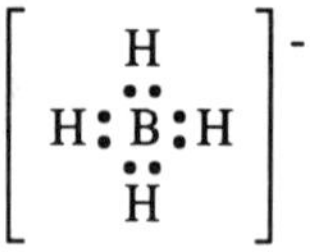 (f) 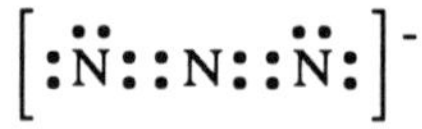 (g) 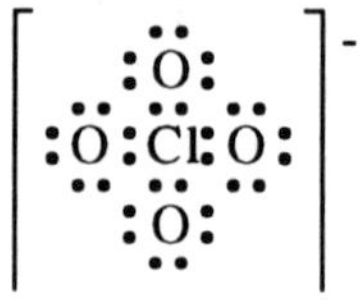 (h) 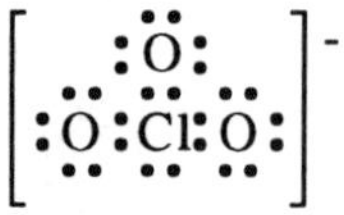

8.9 (a) C (b) O (c) O (d) Mg (e) P (f) Sb

8.10   *Bond*   *Electronegativity difference*
                                    (more electronegative - less electronegative)

(a)   H— Cl   3.0 - 2.1 = 0.9

(b)   Al— H   2.1 - 1.5 = 0.6

(c)   Mg— F   4.0 - 1.2 = 2.8

(d)   O— H   3.5 - 2.1 = 1.4

(e)   Be— S   2.5 - 1.5 = 1.0

(f)   P— S   2.5 - 2.1 = 0.4

(g)   Li— H   2.1 - 1.0 = 1.1

(h) Hydrogen is the *less* electronegative element.

(i) Hydrogen is the *more* electronegative element.

8.11 In order of increasing polarity (electronegativity differences in parentheses):
F–F (0) < C–H (0.4) < C–N (0.5) < H–Cl (0.9) < Al–S (1.0) < Si–O (1.7) < Li–Cl (2.0)

8.12 (a) covalent (b) covalent (c) ionic (d) ionic (e) covalent (f) covalent (g) ionic
(h) covalent

8.13 *Valence electrons* are those electrons beyond the last noble gas core. In the process of
bonding, atoms achieve noble gas-like configurations.

8.14 (a) Al· 3   (b) B· 3   (c) Be· 2   (d) ·C· 4

(e) :Cl· 7   (f) :F· 7   (g) Li· 1   (h) Mg· 2

(i) :N· 5   (j) :P· 5   (k) :S: 6   (l) ·Si· 4

8.15 (a) Li· + ·F̈: → Li⁺ + [:F̈:]⁻

(b) Na·⌒
Na· + ·N̈: → Na⁺ + [:N̈:]³⁻
Na·⌒
Na⁺
Na⁺

(c) Li·⌒
Li· + :Ö: → Li⁺ / Li⁺ + [:Ö:]²⁻

(d) Mg·⌒ ·N̈:
Mg· + → Mg²⁺ + [:N̈:]³⁻
Mg·⌒ ·N̈:
Mg²⁺ [:N̈:]³⁻
Mg²⁺

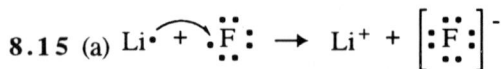

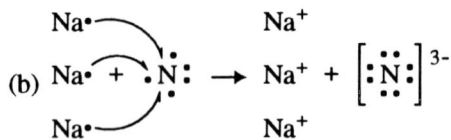

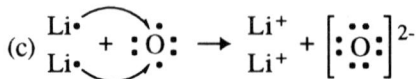

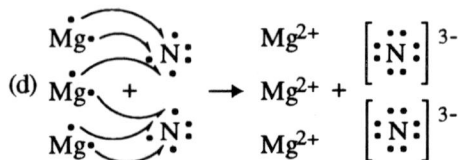

8.16 The bond between a metal and a nonmetal is ionic. The bond between two nonmetals is covalent.

8.17 Refer to Figure 8-4a.

8.18 Refer to Figure 8-5. The region of maximum electron density is between the nuclei.

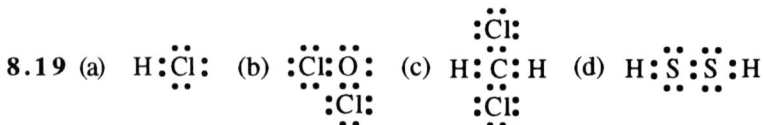

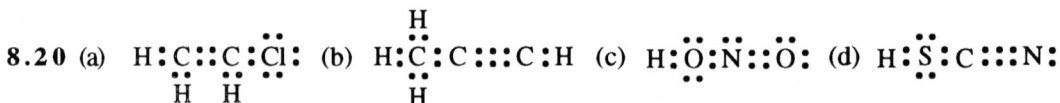

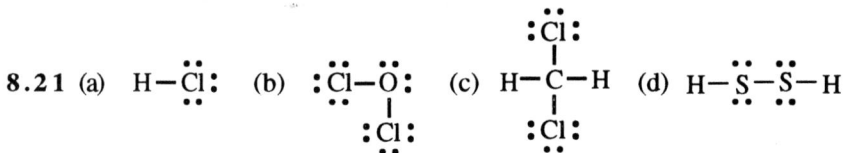

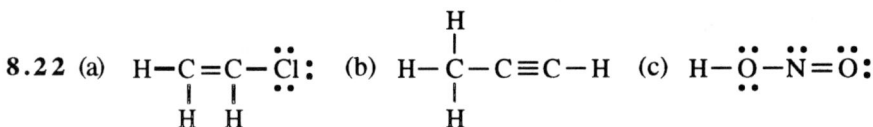

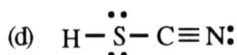

8.19 (a) H:C̈l: (b) :C̈l:Ö:C̈l: (c) H:C̈:H with Cl above and below   (d) H:S̈:S̈:H

8.20 (a) H:C::C:C̈l: with H, H below   (b) H:C:C:::C:H with H above and H, H below   (c) H:Ö:N::Ö: (d) H:S̈:C:::N:

8.21 (a) H—C̈l: (b) :C̈l—Ö—C̈l: with Cl below   (c) H—C—H with :Cl: above and :Cl: below   (d) H—S̈—S̈—H

8.22 (a) H—C=C—C̈l: with H, H below   (b) H—C—C≡C—H with H above and H, H below   (c) H—Ö—N=Ö:

(d) H—S̈—C≡N:

8.23 (a) $\begin{bmatrix} \text{H} \\ \ddot{\text{H}} \\ \text{H:P:H} \\ \text{H} \end{bmatrix}^{+}$ (b) $\begin{bmatrix} :\ddot{\text{Cl}}:\ddot{\text{O}}: \end{bmatrix}^{-}$ (c) $\begin{bmatrix} :\ddot{\text{O}}::\text{N}:\ddot{\text{O}}: \end{bmatrix}^{-}$ (d) $\begin{bmatrix} :\text{O} \quad \text{O}: \\ :\ddot{\text{O}}:\text{C}:\text{C}:\ddot{\text{O}}: \end{bmatrix}^{2-}$

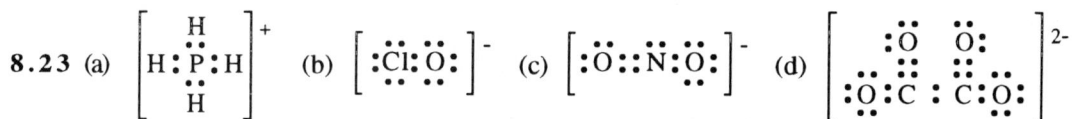

8.24 *Electronegativity* is the relative tendency of an atom to attract a bonded pair of electrons.
(a) Al > Na   (b) Be > Sr   (c) As > Bi   (d) Br > Ge   (e) F > Fr   (f) O > Si

8.25 In order of increasing polarity (electronegativity difference shown in parentheses):
Cl–Cl (0) < N–H (0.9) < C–O (1.0) < Na–H (1.2) < P–F (1.9) < Na–Cl (2.1) < Mg–O
(2.3)

8.26 (a)   covalent   (b) ionic   (c) ionic   (d) covalent (e) covalent   (f) covalent   (g) ionic
(h)   covalent

8.27 [For each of the following answers, the electronegativity difference is in parentheses.]
(a)   nonpolar covalent (0.1)   (b)   polar covalent (0.6)
(c)   polar covalent (1.5)   (d)   nonpolar covalent (0)
(e)   polar covalent (1.9)   (f)   ionic (2.5)
(g)   polar covalent (1.3)   (h)   ionic (3.1)
(i)   polar covalent (0.9)   (j)   nonpolar covalent (0.4)
(k)   It is considered a nonpolar bond.

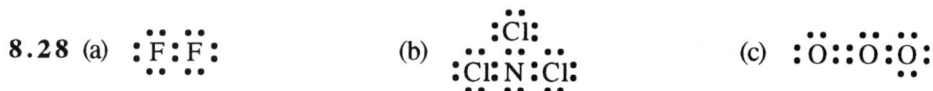

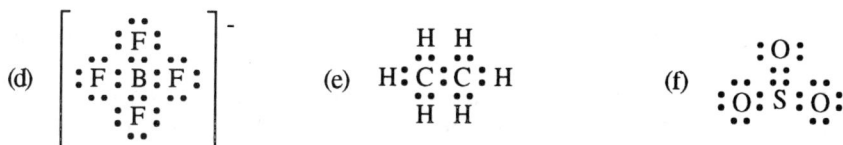

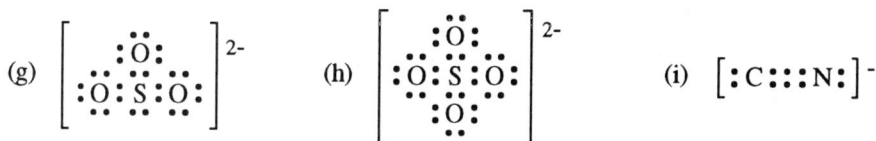

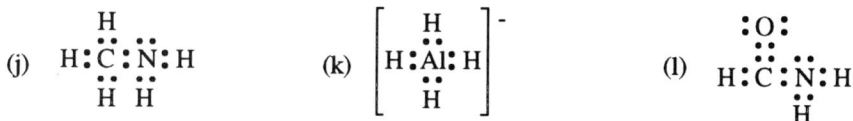

8.28 (a) $:\ddot{\text{F}}:\ddot{\text{F}}:$ (b) $\begin{matrix} :\ddot{\text{Cl}}: \\ :\ddot{\text{Cl}}:\text{N}:\ddot{\text{Cl}}: \end{matrix}$ (c) $:\ddot{\text{O}}::\ddot{\text{O}}:\ddot{\text{O}}:$

(d) $\begin{bmatrix} :\ddot{\text{F}}: \\ :\ddot{\text{F}}:\text{B}:\ddot{\text{F}}: \\ :\ddot{\text{F}}: \end{bmatrix}^{-}$ (e) $\begin{matrix} \text{H} \quad \text{H} \\ \text{H}:\text{C}:\text{C}:\text{H} \\ \text{H} \quad \text{H} \end{matrix}$ (f) $\begin{matrix} :\ddot{\text{O}}: \\ :\ddot{\text{O}}:\text{S}:\ddot{\text{O}}: \end{matrix}$

(g) $\begin{bmatrix} :\ddot{\text{O}}: \\ :\ddot{\text{O}}:\text{S}:\ddot{\text{O}}: \end{bmatrix}^{2-}$ (h) $\begin{bmatrix} :\ddot{\text{O}}: \\ :\ddot{\text{O}}:\text{S}:\ddot{\text{O}}: \\ :\ddot{\text{O}}: \end{bmatrix}^{2-}$ (i) $\begin{bmatrix} :\text{C}:::\text{N}: \end{bmatrix}^{-}$

(j) $\begin{matrix} \text{H} \\ \ddot{\text{H}} \\ \text{H}:\text{C}:\text{N}:\text{H} \\ \text{H} \quad \text{H} \end{matrix}$ (k) $\begin{bmatrix} \text{H} \\ \text{H}:\ddot{\text{Al}}:\text{H} \\ \text{H} \end{bmatrix}^{-}$ (l) $\begin{matrix} :\ddot{\text{O}}: \\ \text{H}:\text{C}:\text{N}:\text{H} \\ \ddot{\text{H}} \end{matrix}$

8.29 (a) Br = $7e^{-}$ Te = $6e^{-}$ Fr = $1e^{-}$ Sn = $4e^{-}$ Ga = $3e^{-}$ Bi = $5e^{-}$
(b) The Roman numeral tells the number of valence electrons.

8.29 (continued):

(c) 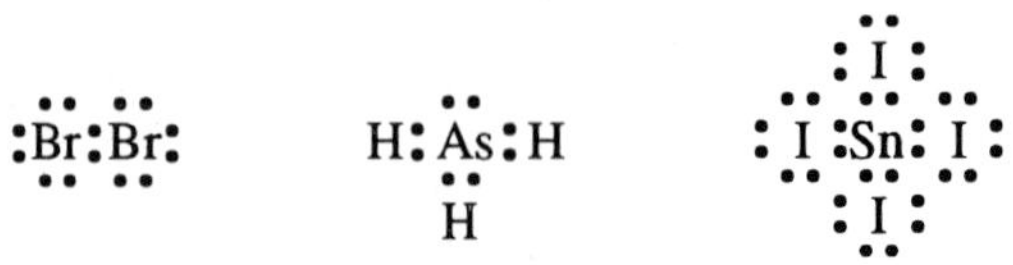   :Br:Br: H:As:H   :I:
      ·· ··   ··   ·· ·· ··
                           H   :I:Sn:I:
                                       ·· ·· ··
                                         :I:
                                          ··

8.30 (a) Element 117 will be a halogen.
   (b) Two atoms of element 117 would form a covalent bond.
   (c) Element 117 would form an ionic bond with cesium (an alkali metal).
   (d) Element 117 would form a covalent bond with iodine (another halogen).
   (e) Element 117 would form a more polar bond with chlorine, because its electronegativity
        difference will be greater with chlorine than with bromine.

8.31 Heat absorbed: 436 kJ + 243 kJ   = 679 kJ
     Heat released: 2×432 kJ   = 864 kJ
     Net heat released: = 185 kJ

8.32 (a) Since the dipoles are equal and in opposite directions along a straight
        line, they cancel one another out.
   (b) Here the dipoles are not pointed in opposite directions and therefore
        do not cancel one another completely. Consequently, there is a net
        dipole in the direction of oxygen through the center of the molecule,
        as shown in the diagram on the right. 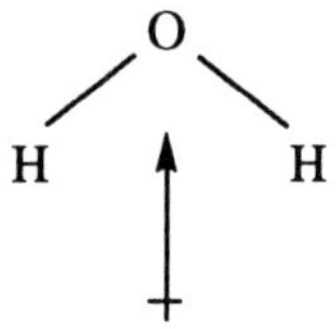
   (c) Boron trifluoride has a zero dipole for a reason similar to that of carbon dioxide, where
        the dipoles cancel one another. In the boron trifluoride case, the dipoles are exactly
        balanced in three directions (rather than two) and cancel each other out in a similar
        fashion.

8.33 (a) The greater the electronegativity *difference,* the greater the magnitude of the partial
        charges.
   (b) Yes. There must be a conservation of charge (the partial charges must add up to zero).
        Thus, the magnitude of the plus charge must equal that of the negative charge.
   (c) The greater the attraction of the electrons to the more electronegative atom, the greater
        the partial charges created.

8.34 (a) CO and CN^- (b) NO_3^- and CO_3^{2-} (c) SO_3^{2-} and PO_3^{3-} (d) NCl_3 and CCl_3^-
   (e) H_3O^+ and NH_3

CHAPTER 9

Chemical Nomenclature

9.1

	Cl^-	O^{2-}	N^{3-}
Na^+	NaCl	Na_2O	Na_3N
Ca^{2+}	$CaCl_2$	CaO	Ca_3N_2
Al^{3+}	$AlCl_3$	Al_2O_3	AlN

9.2 (a) NaCl   (b) KI   (c) $MgBr_2$ (d) BaO
       (e) Li_2O (f) K_2S (g) Al_2S_3 (h) CaF_2
       (i) SrO

9.3 (a) $AlPO_4$ (b) Li_2SO_4 (c) $Al(OH)_3$ (d) $Ba(NO_3)_2$
       (e) $(NH_4)_2CO_3$ (f) $CaSO_4$ (g) $Al_2(CrO_4)_3$ (h) $Sr_3(PO_4)_2$
       (i) $Mg_3(AsO_4)_2$ (j) $Na_2C_2O_4$

9.4 (a) CuBr   (b) $SnCl_2$ (c) $Pb(NO_3)_2$ (d) Co_2O_3
       (e) $Fe(OH)_3$ (f) $Cr_2(SO_4)_3$ (g) Cu_2CO_3 (h) $CoCrO_4$
       (i) HgF_2 (j) Hg_2F_2

9.5 An ionic compound is named by stating the name of the cation followed by the name of the anion.

9.6 (a) barium sulfide   (b) calcium chloride   (c) silver nitride
       (d) copper(II) fluoride   (e) mercury(II) oxide   (f) mercury(I) oxide
       (g) cobalt(II) chloride   (h) cobalt(III) chloride   (i) iron(II) sulfide
       (j) chromium(II) nitride   (k) zinc phosphide   (l) magnesium hydride

9.7 (a) mercury(II) chloride, $HgCl_2$ (b) cobalt(II) iodide, CoI_2
       (c) tin(II) bromide, $SnBr_2$ (d) iron(II) phosphide, Fe_3P_2
       (e) mercury(I) iodide, Hg_2I_2 (f) cobalt(III) nitride, CoN
       (g) lead(II) sulfide, PbS   (h) tin(IV) hydride, SnH_4

9.8 (a) potassium hydroxide   (b) copper(II) cyanide
       (c) aluminum hydroxide   (d) sodium cyanide
       (e) ammonium cyanide   (f) ammonium sulfide

9.9 (a) BrO^- (b) IO_4^- (c) IO_2^- (d) IO^-

9.10 (a) copper(I) sulfate   (b) potassium nitrite   (c) lithium chlorate
       (d) iron(II) sulfite   (e) magnesium nitrate   (f) sodium chlorite

9.11 (a) sodium perchlorate   (b) potassium hypobromite
       (c) calcium hypoiodite   (d) sodium peroxide

9.11 (continued)
   (e) potassium permanganate   (f) silver hypochlorite

9.12 (a) potassium hydrogen sulfate   (b) lithium dihydrogen phosphate
   (c) magnesium hydrogen sulfate   (d) calcium hydrogen carbonate
   (e) magnesium monohydrogen phosphate

9.13 (a) barium chloride dihydrate or barium chloride 2-hydrate
   (b) zinc sulfate heptahydrate or zinc sulfate 7-hydrate
   (c) copper(II) sulfate pentahydrate or copper(II) sulfate 5-hydrate
   (d) cobalt(II) chloride dihydrate or cobalt(II) chloride 2-hydrate

9.14 (a) CO_2 (b) PCl_3 (c) BCl_3 (d) N_2O_4 (e) $SeCl_2$
   (f) SO_2 (g) As_2O_5 (h) NI_3 (i) NO_2 (j) XeF_4

9.15 (a) selenium trioxide   (b) oxygen dichloride
   (c) diphosphorus pentoxide   (d) nitrogen monoxide
   (e) tellurium trioxide   (f) silicon tetrabromide
   (g) nitrogen trifluoride   (h) dinitrogen monoxide
   (i) xenon tetrafluoride   (j) sulfur hexafluoride

9.16 (a) nitric acid   (b) phosphorous acid
   (c) arsenic acid   (d) chromic acid
   (e) nitrous acid   (f) acetic acid
   (g) hydriodic acid   (h) hydrochloric acid
   (i) hydrocyanic acid   (j) aluminum hydroxide
   (k) lithium hydroxide   (l) magnesium hydroxide

9.17 A common name is one that has no particular systematic derivation. A systematic name may be derived from the formula of a substance.

9.18 (a) 7   (b) 4   (c) 8   (d) 5   (e) 1   (f) 6   (g) 2   (h) 3

9.19 (a) KCl (b) MgS (c) BaI_2 (d) Na_2O (e) $AlBr_3$ (f) SrF_2 (g) Al_2O_3

9.20

HCl	HNO_3	H_2CO_3	H_2SO_4	H_3PO_4
$AgCl$	$AgNO_3$	Ag_2CO_3	Ag_2SO_4	Ag_3PO_4
$MgCl_2$	$Mg(NO_3)_2$	$MgCO_3$	$MgSO_4$	$Mg_3(PO_4)_2$
$AlCl_3$	$Al(NO_3)_3$	$Al_2(CO_3)_3$	$Al_2(SO_4)_3$	$AlPO_4$
NH_4Cl	NH_4NO_3	$(NH_4)_2CO_3$	$(NH_4)_2SO_4$	$(NH_4)_3PO_4$

9.21 (a) K_2CO_3 (b) $SrSO_4$ (c) $Mg(OH)_2$ (d) $Al(NO_3)_3$ (e) $Al_2(C_2O_4)_3$
   (f) Na_2CrO_4 (g) $(NH_4)_3PO_4$ (h) $Al_2(CO_3)_3$

9.22 (a) CuI_2 (b) Co_2S_3 (c) $Fe(C_2H_3O_2)_2$ (d) $CoCO_3$ (e) $Cr_2(SO_4)_3$
   (f) $SnCl_4$ (g) SnO_2 (h) $Cu_3(PO_4)_2$

9.23 (a) cobalt(II) sulfide   (b) lithium hydride   (c) copper(II) iodide
(d) iron(III) sulfide   (e) zinc nitride   (f) lead(IV) oxide

9.24 (a) copper(I) chloride, CuCl   (b) cobalt(II) hydroxide, $Co(OH)_2$
(c) chromium(III) oxide, Cr_2O_3 (d) lead(II) nitrate, $Pb(NO_3)_2$
(e) iron(III) sulfate, $Fe_2(SO_4)_3$ (f) tin(IV) chromate, $Sn(CrO_4)_2$

9.25 (a) sodium cyanide   (b) magnesium hydroxide
(c) ammonium fluoride   (d) copper(I) cyanide

9.26 (a) copper(I) sulfate   (b) iron(III) carbonate
(c) magnesium chlorite   (d) chromium(II) arsenate
(e) sodium hydrogen carbonate   (f) copper(II) nitrite
(g) zinc dihydrogen phosphate   (h) calcium acetate
(i) sodium oxalate

9.27 (a) sodium bromate   (b) potassium iodite
(c) ammonium hypochlorite   (d) lithium periodate

9.28 (a) $NaHSO_3$ (b) $LiHCO_3$ (c) $Mg(HCO_3)_2$ (d) $CaHPO_4$
(e) $Ca(H_2PO_4)_2$

9.29 (a) calcium sulfate 2-hydrate (or calcium sulfate dihydrate)
(b) sodium sulfate 10-hydrate (or sodium sulfate decahydrate)
(c) calcium carbonate 6-hydrate (or calcium carbonate hexahydrate)
(d) calcium hypochlorite 3-hydrate (or calcium hypochlorite trihydrate)
(e) mercury(II) nitrate 2-hydrate (or mercury(II) nitrate dihydrate)
(f) cobalt(II) acetate 4-hydrate (or cobalt(II) acetate tetrahydrate)
(g) aluminum oxalate 3-hydrate (or aluminum oxalate trihydrate)
(h) barium peroxide 8-hydrate (or barium peroxide octahydrate)

9.30 (a) N_2S_5 (b) BCl_3 (c) OF_2 (d) SiF_4 (e) N_2O

9.31 (a) phosphorus pentachloride   (b) boron trifluoride
(c) dinitrogen tetroxide   (d) sulfur trioxide
(e) silicon dioxide

9.32 (a) hydrobromic acid   (b) sulfuric acid
(c) sulfurous acid   (d) phosphoric acid
(e) chlorous acid   (f) periodic acid
(g) bromic acid   (h) hydrofluoric acid
(i) hypophosphorous acid   (j) oxalic acid 2-hydrate
(or oxalic acid dihydrate)

9.33 (a) $AgC_2H_3O_2$ (b) Cu_3PO_4 (c) $Hg(NO_3)_2$ (d) $MnSO_4$
(e) $(NH_4)_2Cr_2O_7$ (f) $CdCl_2$ (g) $Sr_3(PO_4)_2$ (h) NH_4HCO_3
(i) K_2HPO_4 (j) KH_2PO_4 (k) SnH_4 (l) Cr_2O_3

9.33 (continued)

(m) PbO_2 (n) Ag_3AsO_4 (o) $Na_2C_2O_4$ (p) MgH_2

(q) Ni_3P_2

9.34 (a) lithium bromide   (b) mercury(I) oxide
(c) potassium permanganate   (d) sodium thiosulfate
(e) nickel(II) sulfite   (f) strontium nitride
(g) ammonium phosphite   (h) sodium peroxide
(i) tin(II) acetate   (j) lead(II) cyanide
(k) silver sulfate   (l) cobalt(II) chromate
(m) magnesium monohydrogen phosphate   (n) manganese(II) arsenate

9.35 (a) $NaHCO_3$, sodium hydrogen carbonate   (b) $NaHSO_3$, sodium hydrogen sulfite
(c) MnO_2, manganese(IV) oxide   (d) SnF_2, tin(II) fluoride
(e) UF_6, uranium(VI) fluoride

9.36 (a) hydrogen selenide   (b) hydroselenic acid
(c) hydrogen telluride   (d) hydrotelluric acid

9.37 (a) gold(I) bromide and gold(III) bromide
(b) antimony(III) fluoride and antimony(V) fluoride
(c) cerium(IV) oxide and cerium(III) oxide
(d) gallium(II) selenide and gallium(III) selenide

9.38 $2 NO_2 \rightarrow N_2O_4$

9.39 (a) hypofluorous acid
(b) HFO_2 is fluorous acid, HFO_3 is fluoric acid, and HFO_4 is perfluoric acid.

9.40 (a) N is +5 in NO_3^- and +3 in NO_2^-.
(b) S is +6 in SO_4^{2-} and +4 in SO_3^{2-}.
(c) P is +5 in PO_4^{3-} and +3 in PO_3^{3-}.
(d) The non-oxygen atom always has a higher oxidation state in the *-ate* ion than it does in the *-ite* ion.
(e) The non-oxygen atom always has a higher oxidation state in the *-ic acid* than it does in the *-ous acid*.

CHAPTER 10

Chemical Changes and Stoichiometry

10.1 A chemical equation is balanced when every atom appearing on the reactant side also appears on the product side, and vice versa. This implies that all of the matter that enters into the reaction as reactants is found somewhere in the products. Hence matter is neither created nor destroyed.

10.2 (a) $C_5H_{12} + 8\,O_2 \longrightarrow 5\,CO_2 + 6\,H_2O$ (b) $H_2 + I_2 \longrightarrow 2\,HI$

(c) $2\,Li + H_2 \longrightarrow 2\,LiH$ (d) $S + O_2 \longrightarrow SO_2$

(e) $2\,Al + Fe_2O_3 \longrightarrow 2\,Fe + Al_2O_3$ (f) $2\,Fe_2S_3 + 9\,O_2 \longrightarrow 2\,Fe_2O_3 + 6\,SO_2$

(g) $2\,KOH + H_2SO_4 \longrightarrow K_2SO_4 + 2\,H_2O$ (h) $P_4 + 5\,O_2 \longrightarrow P_4O_{10}$

(i) $Zn + 2\,HCl \longrightarrow ZnCl_2 + H_2$ (j) $2\,Al + 3\,H_2SO_4 \longrightarrow Al_2(SO_4)_3 + 3\,H_2$

(k) $C_6H_{12}O_6 + 6\,O_2 \longrightarrow 6\,CO_2 + 6\,H_2O$ (l) $Mg_3N_2 + 3\,H_2O \longrightarrow 3\,MgO + 2\,NH_3$

(m) $N_2H_4 + 3\,O_2 \longrightarrow 2\,NO_2 + 2\,H_2O$

10.3 (a) pentane + oxygen $\longrightarrow$ carbon dioxide + water

$?\,C_5H_{12} + ?\,O_2 \longrightarrow ?\,CO_2 + ?\,H_2O$

$C_5H_{12} + 8\,O_2 \longrightarrow 5\,CO_2 + 6\,H_2O$

(b) mercury(II) oxide $\longrightarrow$ mercury + oxygen

$?\,HgO \longrightarrow ?\,Hg + ?\,O_2$

$2\,HgO \longrightarrow 2\,Hg + O_2$

(c) phosphoric acid + potassium hydroxide $\longrightarrow$ potassium phosphate + water

$?\,H_3PO_4 + ?\,KOH \longrightarrow ?\,K_3PO_4 + ?\,H_2O$

$H_3PO_4 + 3\,KOH \longrightarrow K_3PO_4 + 3\,H_2O$

(d) magnesium carbonate + nitric acid $\longrightarrow$ magnesium nitrate + water + carbon dioxide

$?\,MgCO_3 + ?\,HNO_3 \longrightarrow ?\,Mg(NO_3)_2 + ?\,H_2O + ?\,CO_2$

$MgCO_3 + 2\,HNO_3 \longrightarrow Mg(NO_3)_2 + H_2O + CO_2$

10.4 (a) magnesium + oxygen $\longrightarrow$ magnesium oxide

$?\,Mg + ?\,O_2 \longrightarrow ?\,MgO$

$2\,Mg + O_2 \longrightarrow 2\,MgO$

(b) barium oxide + hydrochloric acid $\longrightarrow$ barium chloride + water

$?\,BaO + ?\,HCl \longrightarrow ?\,BaCl_2 + ?\,H_2O$

$BaO + 2\,HCl \longrightarrow BaCl_2 + H_2O$

10.4 (continued):

(c) sulfur trioxide + water $\longrightarrow$ sulfuric acid

? SO_3 + ? H_2O $\longrightarrow$? H_2SO_4

SO_3 + H_2O $\longrightarrow$ H_2SO_4

(d) sodium + water $\longrightarrow$ sodium hydroxide + hydrogen

? Na + ? H_2O $\longrightarrow$? $NaOH$ + ? H_2

$2\,Na$ + $2\,H_2O$ $\longrightarrow$ $2\,NaOH$ + H_2

(e) sodium carbonate + sulfuric acid $\longrightarrow$ sodium sulfate + carbon dioxide + water

? Na_2CO_3 + ? H_2SO_4 $\longrightarrow$? Na_2SO_4 + ? CO_2 + ? H_2O

Na_2CO_3 + H_2SO_4 $\longrightarrow$ Na_2SO_4 + CO_2 + H_2O

(f) diphosphorus pentoxide + water $\longrightarrow$ phosphoric acid

? P_2O_5 + ? H_2O $\longrightarrow$? H_3PO_4

P_2O_5 + $3\,H_2O$ $\longrightarrow$ $2\,H_3PO_4$

(g) aluminum + oxygen $\longrightarrow$ aluminum oxide

? Al + ? O_2 $\longrightarrow$? Al_2O_3

$4\,Al$ + $3\,O_2$ $\longrightarrow$ $2\,Al_2O_3$

(h) iron(III) oxide + hydrogen $\longrightarrow$ iron + water

? Fe_2O_3 + ? H_2 $\longrightarrow$? Fe + ? H_2O

Fe_2O_3 + $3\,H_2$ $\longrightarrow$ $2\,Fe$ + $3\,H_2O$

10.5 (a) $C_5H_{12}(l)$ + $8\,O_2(g)$ $\longrightarrow$ $5\,CO_2(g)$ + $6\,H_2O(l)$

(b) $MgCO_3(s)$ + $2\,HNO_3(aq)$ $\longrightarrow$ $Mg(NO_3)_2(aq)$ + $CO_2(g)$ + $H_2O(\ell)$

(c) $2\,Na(s)$ + $2\,H_2O(\ell)$ $\longrightarrow$ $2\,NaOH(aq)$ + $H_2(g)$

(d) $2\,HgO(s)$ $\longrightarrow$ $2\,Hg(l)$ + $O_2(g)$

(e) $2\,C(s)$ + $O_2(g)$ $\longrightarrow$ $2\,CO(g)$

(f) $Pb(NO_3)_2(aq)$ + $2\,KI(aq)$ $\longrightarrow$ $PbI_2(s)$ + $2\,KNO_3(aq)$

10.6 (a) 3 mol Cu : 8 mol HNO_3

(b) 3 mol Cu : 3 mol $Cu(NO_3)_2$ [or 1 mol Cu : 1 mol $Cu(NO_3)_2$]

(c) 8 mol HNO_3 : 2 mol NO [or 4 mol HNO_3 : 1 mol NO]

(d) 2 mol NO : 4 mol H_2O [or 1 mol NO : 2 mol H_2O]

(e) 3 mol Cu : 2 mol NO

(f) 3 mol $Cu(NO_3)_2$: 4 mol H_2O

10.7 (a) C_5H_{12} + $8\,O_2$ $\longrightarrow$ $5\,CO_2$ + $6\,H_2O$

(b) $2.10 \text{ mol } C_5H_{12} \left(\dfrac{8 \text{ mol } O_2}{1 \text{ mol } C_5H_{12}} \right) = 16.8 \text{ mol } O_2$

10.7 (continued):

(c) $2.10 \text{ mol } C_5H_{12} \left(\dfrac{5 \text{ mol } CO_2}{1 \text{ mol } C_5H_{12}} \right) = 10.5 \text{ mol } CO_2$

(d) $2.10 \text{ mol } C_5H_{12} \left(\dfrac{6 \text{ mol } H_2O}{1 \text{ mol } C_5H_{12}} \right) = 12.6 \text{ mol } H_2O$

10.8 (a) $2 \text{ Al} + 6 \text{ HCl} \longrightarrow 2 \text{ AlCl}_3 + 3 \text{ H}_2$

(b) $0.450 \text{ mol Al} \left(\dfrac{6 \text{ mol HCl}}{2 \text{ mol Al}} \right) = 1.35 \text{ mol HCl}$

(c) $0.450 \text{ mol Al} \left(\dfrac{2 \text{ mol AlCl}_3}{2 \text{ mol Al}} \right) = 0.450 \text{ mol AlCl}_3$

(d) $0.450 \text{ mol Al} \left(\dfrac{3 \text{ mol } H_2}{2 \text{ mol Al}} \right) = 0.675 \text{ mol } H_2$

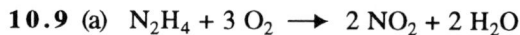

10.9 (a) $N_2H_4 + 3 O_2 \longrightarrow 2 NO_2 + 2 H_2O$

(b) $1.30 \text{ mol } N_2H_4 \left(\dfrac{3 \text{ mol } O_2}{1 \text{ mol } N_2H_4} \right) = 3.90 \text{ mol } O_2$

(c) $1.30 \text{ mol } N_2H_4 \left(\dfrac{2 \text{ mol } NO_2}{1 \text{ mol } N_2H_4} \right) = 2.60 \text{ mol } NO_2$

(d) $1.30 \text{ mol } N_2H_4 \left(\dfrac{2 \text{ mol } H_2O}{1 \text{ mol } N_2H_4} \right) = 2.60 \text{ mol } H_2O$

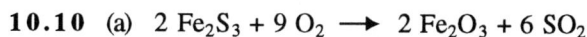

10.10 (a) $2 \text{ Fe}_2S_3 + 9 O_2 \longrightarrow 2 \text{ Fe}_2O_3 + 6 SO_2$

(b) $0.900 \text{ mol } SO_2 \left(\dfrac{2 \text{ mol Fe}_2S_3}{6 \text{ mol } SO_2} \right) = 0.300 \text{ mol Fe}_2S_3$

(c) $0.900 \text{ mol } SO_2 \left(\dfrac{9 \text{ mol } O_2}{6 \text{ mol } SO_2} \right) = 1.35 \text{ mol } O_2$

(d) $0.900 \text{ mol } SO_2 \left(\dfrac{2 \text{ mol Fe}_2O_3}{6 \text{ mol } SO_2} \right) = 0.300 \text{ mol Fe}_2O_3$

10.11 (a) $14.0 \text{ g CO} \left(\dfrac{1 \text{ mol CO}}{28.0 \text{ g CO}} \right) = 0.500 \text{ mol CO}$

10.11 (continued):

(b) $25.0 \text{ g CaCO}_3 \left(\dfrac{1 \text{ mol CaCO}_3}{100.1 \text{ g CaCO}_3} \right) = 0.250 \text{ mol CaCO}_3$

(c) $30.0 \text{ g Na}_2\text{SO}_4 \left(\dfrac{1 \text{ mol Na}_2\text{SO}_4}{142.1 \text{ g Na}_2\text{SO}_4} \right) = 0.211 \text{ mol Na}_2\text{SO}_4$

(d) $62.9 \text{ g (NH}_4)_3\text{PO}_4 \left(\dfrac{1 \text{ mol (NH}_4)_3\text{PO}_4}{149.0 \text{ g (NH}_4)_3\text{PO}_4} \right) = 0.422 \text{ mol (NH}_4)_3\text{PO}_4$

10.12   (a) $0.200 \text{ mol CO}_2 \left(\dfrac{44.0 \text{ g CO}_2}{1 \text{ mol CO}_2} \right) = 88.0 \text{ g CO}_2$

(b) $0.300 \text{ mol SO}_3 \left(\dfrac{80.1 \text{ g SO}_3}{1 \text{ mol SO}_3} \right) = 24.0 \text{ g SO}_3$

(c) $0.835 \text{ mol C}_7\text{H}_{16} \left(\dfrac{100.0 \text{ g C}_7\text{H}_{16}}{1 \text{ mol C}_7\text{H}_{16}} \right) = 83.5 \text{ g C}_7\text{H}_{16}$

(d) $0.0175 \text{ mol Ca(OH)}_2 \left(\dfrac{74.1 \text{ g Ca(OH)}_2}{1 \text{ mol Ca(OH)}_2} \right) = 1.30 \text{ g Ca(OH)}_2$

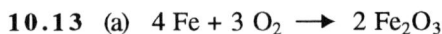

10.13   (a) $4 \text{ Fe} + 3 \text{ O}_2 \longrightarrow 2 \text{ Fe}_2\text{O}_3$

(b) $5.58 \text{ g Fe} \left(\dfrac{1 \text{ mol Fe}}{55.8 \text{ g Fe}} \right) \left(\dfrac{2 \text{ mol Fe}_2\text{O}_3}{4 \text{ mol Fe}} \right) \left(\dfrac{159.6 \text{ g Fe}_2\text{O}_3}{1 \text{ mol Fe}_2\text{O}_3} \right) = 7.98 \text{ g Fe}_2\text{O}_3$

10.14   (a) $\text{C}_6\text{H}_{12}\text{O}_6 + 6 \text{ O}_2 \longrightarrow 6 \text{ CO}_2 + 6 \text{ H}_2\text{O}$

(b) $36.0 \text{ g C}_6\text{H}_{12}\text{O}_6 \left(\dfrac{1 \text{ mol C}_6\text{H}_{12}\text{O}_6}{180.0 \text{ g C}_6\text{H}_{12}\text{O}_6} \right) = 0.200 \text{ mol C}_6\text{H}_{12}\text{O}_6$

$0.200 \text{ mol C}_6\text{H}_{12}\text{O}_6 \left(\dfrac{6 \text{ mol O}_2}{1 \text{ mol C}_6\text{H}_{12}\text{O}_6} \right) \left(\dfrac{32.0 \text{ g O}_2}{1 \text{ mol O}_2} \right) = 38.4 \text{ g O}_2$

(c) $0.200 \text{ mol C}_6\text{H}_{12}\text{O}_6 \left(\dfrac{6 \text{ mol CO}_2}{1 \text{ mol C}_6\text{H}_{12}\text{O}_6} \right) \left(\dfrac{44.0 \text{ g CO}_2}{1 \text{ mol CO}_2} \right) = 52.8 \text{ g CO}_2$

(d) $0.200 \text{ mol C}_6\text{H}_{12}\text{O}_6 \left(\dfrac{6 \text{ mol H}_2\text{O}}{1 \text{ mol C}_6\text{H}_{12}\text{O}_6} \right) \left(\dfrac{18.0 \text{ g H}_2\text{O}}{1 \text{ mol H}_2\text{O}} \right) = 21.6 \text{ g H}_2\text{O}$

(e) $36.0 \text{ g} + 38.4 \text{ g} = 74.4 \text{ g} = 52.8 \text{ g} + 21.6 \text{ g}$

10.15 (a) $N_2 + 3 H_2 \longrightarrow 2 NH_3$

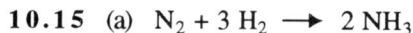

(b) $85.0 \text{ g } NH_3 \left(\dfrac{1 \text{ mol } NH_3}{17.0 \text{ g } NH_3} \right) = 5.00 \text{ mol } NH_3$

$5.00 \text{ mol } NH_3 \left(\dfrac{3 \text{ mol } H_2}{2 \text{ mol } NH_3} \right) \left(\dfrac{2.0 \text{ g } H_2}{1 \text{ mol } H_2} \right) = 15 \text{ g } H_2$

(c) $5.00 \text{ mol } NH_3 \left(\dfrac{1 \text{ mol } N_2}{2 \text{ mol } NH_3} \right) \left(\dfrac{28.0 \text{ g } N_2}{1 \text{ mol } N_2} \right) = 70.0 \text{ g } N_2$

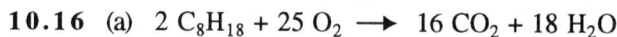

10.16 (a) $2 C_8H_{18} + 25 O_2 \longrightarrow 16 CO_2 + 18 H_2O$

(b) $702 \text{ g } C_8H_{18} \left(\dfrac{1 \text{ mol } C_8H_{18}}{114.0 \text{ g } C_8H_{18}} \right) = 6.16 \text{ mol } C_8H_{18}$

$6.16 \text{ mol } C_8H_{18} \left(\dfrac{16 \text{ mol } CO_2}{2 \text{ mol } C_8H_{18}} \right) \left(\dfrac{44.0 \text{ g } CO_2}{1 \text{ mol } CO_2} \right) = 2170 \text{ g } CO_2 \ (2.17 \times 10^3 \text{ g})$

(c) $6.16 \text{ mol } C_8H_{18} \left(\dfrac{25 \text{ mol } O_2}{2 \text{ mol } C_8H_{18}} \right) \left(\dfrac{32.0 \text{ g } O_2}{1 \text{ mol } O_2} \right) = 2460 \text{ g } O_2 \ (2.46 \times 10^3 \text{ g})$

10.17 As in all stoichiometry problems, we must begin with a balanced equation.
$2 Al + 3 H_2SO_4 \longrightarrow Al_2(SO_4)_3 + 3 H_2$

(a) $455 \text{ g } Al_2(SO_4)_3 \left(\dfrac{1 \text{ mol } Al_2(SO_4)_3}{342.3 \text{ g } Al_2(SO_4)_3} \right) = 1.33 \text{ mol } Al_2(SO_4)_3$

$1.33 \text{ mol } Al_2(SO_4)_3 \left(\dfrac{2 \text{ mol } Al}{1 \text{ mol } Al_2(SO_4)_3} \right) \left(\dfrac{27.0 \text{ g } Al}{1 \text{ mol } Al} \right) = 71.8 \text{ g } Al$

(b) $1.33 \text{ mol } Al_2(SO_4)_3 \left(\dfrac{3 \text{ mol } H_2SO_4}{1 \text{ mol } Al_2(SO_4)_3} \right) \left(\dfrac{98.1 \text{ g } H_2SO_4}{1 \text{ mol } H_2SO_4} \right) = 391 \text{ g } H_2SO_4$

(c) $1.33 \text{ mol } Al_2(SO_4)_3 \left(\dfrac{3 \text{ mol } H_2}{1 \text{ mol } Al_2(SO_4)_3} \right) \left(\dfrac{2.0 \text{ g } H_2}{1 \text{ mol } H_2} \right) = 8.0 \text{ g } H_2$

10.18 Theoretical yield:

$3.28 \text{ g } C_7H_6O_3 \left(\dfrac{1 \text{ mol } C_7H_6O_3}{138.0 \text{ g } C_7H_6O_3} \right) \left(\dfrac{1 \text{ mol } C_9H_8O_4}{1 \text{ mol } C_7H_6O_3} \right) \left(\dfrac{180.0 \text{ g } C_9H_8O_4}{1 \text{ mol } C_9H_8O_4} \right) = 4.28 \text{ g }$
$C_9H_8O_4$

Percentage yield: $\dfrac{3.11 \text{ g}}{4.28 \text{ g}} \times 100\% = 72.7\%$

10.19 Theoretical yield:

$$17.5 \text{ g C}_6\text{H}_6\left(\frac{1 \text{ mol C}_6\text{H}_6}{78.0 \text{ g C}_6\text{H}_6}\right)\left(\frac{1 \text{ mol C}_6\text{H}_5\text{NO}_2}{1 \text{ mol C}_6\text{H}_6}\right)\left(\frac{123.0 \text{ g C}_6\text{H}_5\text{NO}_2}{1 \text{ mol C}_6\text{H}_5\text{NO}_2}\right) = 27.6 \text{ g}$$

$\text{C}_6\text{H}_5\text{NO}_2$

Percentage yield: $\dfrac{22.6 \text{ g}}{27.6 \text{ g}} \times 100\% = 81.9\%$

10.20 Theoretical yield:

$$5.36 \text{ g Pb(NO}_3)_2\left(\frac{1 \text{ mol Pb(NO}_3)_2}{331.2 \text{ g Pb(NO}_3)_2}\right)\left(\frac{1 \text{ mol PbCl}_2}{1 \text{ mol Pb(NO}_3)_2}\right)\left(\frac{278.2 \text{ g PbCl}_2}{1 \text{ mol PbCl}_2}\right)$$

$= 4.50 \text{ g PbCl}_2$

Percentage yield: $\dfrac{3.21 \text{ g}}{4.50 \text{ g}} \times 100\% = 71.3\%$

10.21 $8.50 \text{ g Mg}\left(\dfrac{1 \text{ mol Mg}}{24.3 \text{ g Mg}}\right)\left(\dfrac{1 \text{ mol S}}{1 \text{ mol Mg}}\right)\left(\dfrac{32.1 \text{ g S}}{1 \text{ mol S}}\right) = 11.2 \text{ g S}$

Only 9.00 g S are present, so there is not enough sulfur to react with all of the magnesium. Hence sulfur is the limiting reagent.

$9.00 \text{ g S}\left(\dfrac{1 \text{ mol S}}{32.1 \text{ g S}}\right)\left(\dfrac{1 \text{ mol MgS}}{1 \text{ mol S}}\right)\left(\dfrac{56.4 \text{ g MgS}}{1 \text{ mol MgS}}\right) = 15.8 \text{ g MgS}$

10.22 $5.75 \text{ g BaCl}_2\left(\dfrac{1 \text{ mol BaCl}_2}{208.3 \text{ g BaCl}_2}\right)\left(\dfrac{1 \text{ mol Na}_2\text{SO}_4}{1 \text{ mol BaCl}_2}\right)\left(\dfrac{142.1 \text{ g Na}_2\text{SO}_4}{1 \text{ mol Na}_2\text{SO}_4}\right) = 3.92 \text{ g Na}_2\text{SO}_4$

Because 4.25 g Na_2SO_4 are available, sodium sulfate is in excess and barium chloride is the limiting reagent.

$5.75 \text{ g BaCl}_2\left(\dfrac{1 \text{ mol BaCl}_2}{208.3 \text{ g BaCl}_2}\right)\left(\dfrac{1 \text{ mol BaSO}_4}{1 \text{ mol BaCl}_2}\right)\left(\dfrac{233.4 \text{ g BaSO}_4}{1 \text{ mol BaSO}_4}\right) = 6.44 \text{ g BaSO}_4$

10.23 $4.55 \text{ g Pb(NO}_3)_2\left(\dfrac{1 \text{ mol Pb(NO}_3)_2}{331.2 \text{ g Pb(NO}_3)_2}\right)\left(\dfrac{2 \text{ mol KI}}{1 \text{ mol Pb(NO}_3)_2}\right)\left(\dfrac{166.0 \text{ g KI}}{1 \text{ mol KI}}\right) = 4.56 \text{ g KI}$

Only 3.75 g KI are actually present, so there is not enough potassium iodide to react with all of the lead(II) nitrate. Thus, potassium iodide is the limiting reagent.

$3.75 \text{ g KI}\left(\dfrac{1 \text{ mol KI}}{166.0 \text{ g KI}}\right)\left(\dfrac{1 \text{ mol PbI}_2}{2 \text{ mol KI}}\right)\left(\dfrac{461.0 \text{ g PbI}_2}{1 \text{ mol PbI}_2}\right) = 5.21 \text{ g PbI}_2$

10.24 (a) endothermic (b) exothermic (c) exothermic (d) endothermic

10.25 $50.0 \text{ g HgO} \left(\dfrac{1 \text{ mol HgO}}{216.6 \text{ g HgO}} \right) \left(\dfrac{182 \text{ kJ}}{2 \text{ mol HgO}} \right) = 21.0 \text{ kJ}$

10.26 A *word equation* is a description of a chemical reaction using the names of the reactants and products, rather than their symbols. A *reactant* is a substance that undergoes a change in a chemical reaction. A *product* is a substance formed in a chemical reaction. An *unbalanced equation* is a chemical equation prior to adjusting the coefficients. A *balanced equation* is a chemical equation in which the coefficients have been adjusted so that every atom appearing on the left appears on the right and vice versa. During a chemical reaction the atoms become rearranged.

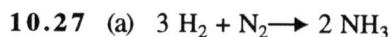

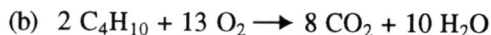

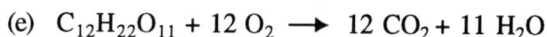

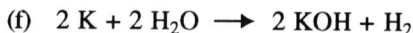

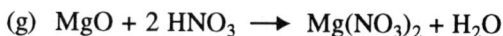

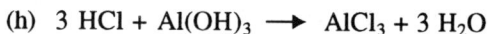

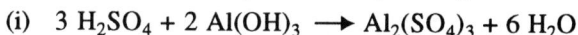

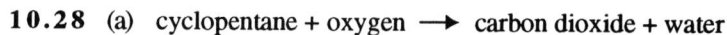

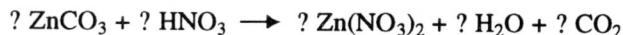

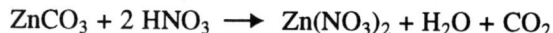

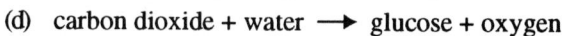

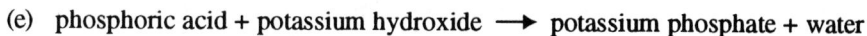

10.27 (a) $3 H_2 + N_2 \longrightarrow 2 NH_3$

(b) $2 C_4H_{10} + 13 O_2 \longrightarrow 8 CO_2 + 10 H_2O$

(c) $Ca + 2 H_2O \longrightarrow Ca(OH)_2 + H_2$

(d) $2 SO_2 + O_2 \longrightarrow 2 SO_3$

(e) $C_{12}H_{22}O_{11} + 12 O_2 \longrightarrow 12 CO_2 + 11 H_2O$

(f) $2 K + 2 H_2O \longrightarrow 2 KOH + H_2$

(g) $MgO + 2 HNO_3 \longrightarrow Mg(NO_3)_2 + H_2O$

(h) $3 HCl + Al(OH)_3 \longrightarrow AlCl_3 + 3 H_2O$

(i) $3 H_2SO_4 + 2 Al(OH)_3 \longrightarrow Al_2(SO_4)_3 + 6 H_2O$

(j) $2 CCl_4 + O_2 \longrightarrow 2 COCl_2 + 2 Cl_2$

(k) $3 KClO \longrightarrow 2 KCl + KClO_3$

(l) $PbO_2 + Pb + 2 H_2SO_4 \longrightarrow 2 PbSO_4 + 2 H_2O$

10.28 (a) cyclopentane + oxygen $\longrightarrow$ carbon dioxide + water

$? C_5H_{10} + ? O_2 \longrightarrow ? CO_2 + ? H_2O$

$2 C_5H_{10} + 15 O_2 \longrightarrow 10 CO_2 + 10 H_2O$

(b) zinc carbonate + nitric acid $\longrightarrow$ zinc nitrate + water + carbon dioxide

$? ZnCO_3 + ? HNO_3 \longrightarrow ? Zn(NO_3)_2 + ? H_2O + ? CO_2$

$ZnCO_3 + 2 HNO_3 \longrightarrow Zn(NO_3)_2 + H_2O + CO_2$

(c) thionyl chloride + water $\longrightarrow$ sulfur dioxide + hydrogen chloride

$? SOCl_2 + ? H_2O \longrightarrow ? SO_2 + ? HCl$

$SOCl_2 + H_2O \longrightarrow SO_2 + 2 HCl$

(d) carbon dioxide + water $\longrightarrow$ glucose + oxygen

$? CO_2 + ? H_2O \longrightarrow ? C_6H_{12}O_6 + ? O_2$

$6 CO_2 + 6 H_2O \longrightarrow C_6H_{12}O_6 + 6 O_2$

(e) phosphoric acid + potassium hydroxide $\longrightarrow$ potassium phosphate + water

$? H_3PO_4 + ? KOH \longrightarrow ? K_3PO_4 + ? H_2O$

$H_3PO_4 + 3 KOH \longrightarrow K_3PO_4 + 3 H_2O$

10.28 (continued):

    (f) aluminum + hydrobromic acid $\longrightarrow$ aluminum bromide + hydrogen

        ? Al + ? HBr $\longrightarrow$? $AlBr_3$ + ? H_2

        2 Al + 6 HBr $\longrightarrow$ 2 $AlBr_3$ + 3 H_2

    (g) copper + sulfur $\longrightarrow$ copper(I) sulfide

        ? Cu + ? S $\longrightarrow$? Cu_2S

        2 Cu + S $\longrightarrow$ Cu_2S

    (h) potassium chlorate $\longrightarrow$ potassium chloride + oxygen

        ? $KClO_3$ $\longrightarrow$? KCl + ? O_2

        2 $KClO_3$ $\longrightarrow$ 2 KCl + 3 O_2

10.29 (a) reacts to form (b) plus (c) solid (d) liquid (e) gas (f) aqueous

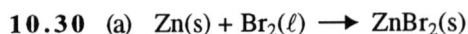

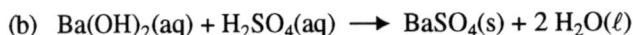

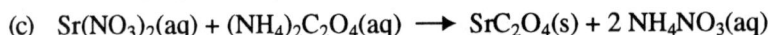

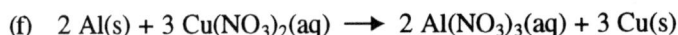

10.30 (a) $Zn(s) + Br_2(\ell) \longrightarrow ZnBr_2(s)$

    (b) $Ba(OH)_2(aq) + H_2SO_4(aq) \longrightarrow BaSO_4(s) + 2\ H_2O(\ell)$

    (c) $Sr(NO_3)_2(aq) + (NH_4)_2C_2O_4(aq) \longrightarrow SrC_2O_4(s) + 2\ NH_4NO_3(aq)$

    (d) $2\ C_4H_{10}(g)\ 13\ O_2(g) \longrightarrow 8\ CO_2(g) + 10\ H_2O(\ell)$

    (e) $CaCO_3(s) + 2\ HCl(aq) \longrightarrow CaCl_2(aq) + CO_2(g) + H_2O(\ell)$

    (f) $2\ Al(s) + 3\ Cu(NO_3)_2(aq) \longrightarrow 2\ Al(NO_3)_3(aq) + 3\ Cu(s)$

10.31 (a) $2.40 \text{ mol } C_3H_5N_3O_9 \left(\dfrac{1 \text{ mol } O_2}{4 \text{ mol } C_3H_5N_3O_9} \right) = 0.600 \text{ mol } O_2$

    (b) $3.60 \text{ mol } C_3H_5N_3O_9 \left(\dfrac{6 \text{ mol } N_2}{4 \text{ mol } C_3H_5N_3O_9} \right) = 5.40 \text{ mol } N_2$

    (c) $5.25 \text{ mol } H_2O \left(\dfrac{4 \text{ mol } C_3H_5N_3O_9}{10 \text{ mol } H_2O} \right) = 2.10 \text{ mol } C_3H_5N_3O_9$

    (d) $1.80 \text{ mol } CO_2 \left(\dfrac{10 \text{ mol } H_2O}{12 \text{ mol } CO_2} \right) = 1.50 \text{ mol } H_2O$

10.32 (a) $4\ NH_3 + 3\ O_2 \longrightarrow 2\ N_2 + 6\ H_2O$

    (b) $0.624 \text{ mol } NH_3 \left(\dfrac{3 \text{ mol } O_2}{4 \text{ mol } NH_3} \right) = 0.468 \text{ mol } O_2$

    (b) $0.306 \text{ mol } H_2O \left(\dfrac{4 \text{ mol } NH_3}{6 \text{ mol } H_2O} \right) = 0.204 \text{ mol } NH_3$

10.32 (continued):

(c) $0.258 \text{ mol NH}_3 \left(\dfrac{2 \text{ mol N}_2}{4 \text{ mol NH}_3} \right) = 0.129 \text{ mol N}_2$

(d) $0.738 \text{ mol H}_2\text{O} \left(\dfrac{2 \text{ mol N}_2}{6 \text{ mol H}_2\text{O}} \right) = 0.246 \text{ mol N}_2$

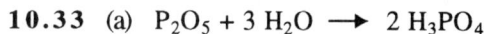

10.33 (a) $P_2O_5 + 3 \text{ H}_2\text{O} \longrightarrow 2 \text{ H}_3\text{PO}_4$

(b) $3.88 \text{ mol H}_3\text{PO}_4 \left(\dfrac{1 \text{ mol P}_2\text{O}_5}{2 \text{ mol H}_3\text{PO}_4} \right) = 1.94 \text{ mol P}_2\text{O}_5$

(c) $3.88 \text{ mol H}_3\text{PO}_4 \left(\dfrac{3 \text{ mol H}_2\text{O}}{2 \text{ mol H}_3\text{PO}_4} \right) = 5.82 \text{ mol H}_2\text{O}$

10.34 (a) $44.0 \text{ g NaOH} \left(\dfrac{1 \text{ mol NaOH}}{40.0 \text{ g NaOH}} \right) = 1.10 \text{ mol NaOH}$

(b) $12.6 \text{ g Al(NO}_3)_3 \left(\dfrac{1 \text{ mol Al(NO}_3)_3}{213.0 \text{ g Al(NO}_3)_3} \right) = 0.0592 \text{ mol Al(NO}_3)_3$

(c) $4.85 \text{ mol H}_2\text{SO}_4 \left(\dfrac{98.1 \text{ g H}_2\text{SO}_4}{1 \text{ mol H}_2\text{SO}_4} \right) = 476 \text{ g H}_2\text{SO}_4$

(d) $0.250 \text{ mol O}_3 \left(\dfrac{48.0 \text{ g O}_3}{1 \text{ mol O}_3} \right) = 12.0 \text{ g O}_3$

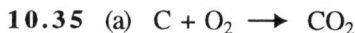

10.35 (a) $C + O_2 \longrightarrow CO_2$

(b) $24.0 \text{ g C} \left(\dfrac{1 \text{ mol C}}{12.0 \text{ g C}} \right) = 2.00 \text{ mol C}$

$2.00 \text{ mol C} \left(\dfrac{1 \text{ mol O}_2}{1 \text{ mol C}} \right)\left(\dfrac{32.0 \text{ g O}_2}{1 \text{ mol O}_2} \right) = 64.0 \text{ g O}_2$

(c) $2.00 \text{ mol C} \left(\dfrac{1 \text{ mol CO}_2}{1 \text{ mol C}} \right)\left(\dfrac{44.0 \text{ g CO}_2}{1 \text{ mol CO}_2} \right) = 88.0 \text{ g CO}_2$

(d) $24.0 \text{ g C} + 64.0 \text{ g O}_2 = 88.0 \text{ g reactants}$
$88.0 \text{ g CO}_2 = 88.0 \text{ g products}$

10.36 $175 \text{ g C}_2\text{H}_2 \left(\dfrac{1 \text{ mol C}_2\text{H}_2}{26.0 \text{ g C}_2\text{H}_2} \right)\left(\dfrac{1 \text{ mol CaC}_2}{1 \text{ mol C}_2\text{H}_2} \right)\left(\dfrac{64.1 \text{ g CaC}_2}{1 \text{ mol CaC}_2} \right) = 431 \text{ g CaC}_2$

10.37 (a) $20.0 \text{ g Cu} \left(\dfrac{1 \text{ mol Cu}}{63.5 \text{ g Cu}} \right) = 0.315 \text{ mol Cu}$

$0.315 \text{ mol Cu} \left(\dfrac{8 \text{ mol HNO}_3}{3 \text{ mol Cu}} \right) \left(\dfrac{63.0 \text{ g HNO}_3}{1 \text{ mol HNO}_3} \right) = 52.9 \text{ g HNO}_3$

(b) $0.315 \text{ mol Cu} \left(\dfrac{2 \text{ mol NO}}{3 \text{ mol Cu}} \right) \left(\dfrac{30.0 \text{ g NO}}{1 \text{ mol NO}} \right) = 6.30 \text{ g NO}$

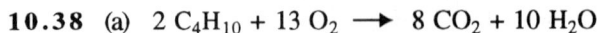

10.38 (a) $2 \text{ C}_4\text{H}_{10} + 13 \text{ O}_2 \longrightarrow 8 \text{ CO}_2 + 10 \text{ H}_2\text{O}$

(b) $196 \text{ g C}_4\text{H}_{10} \left(\dfrac{1 \text{ mol C}_4\text{H}_{10}}{58.0 \text{ g C}_4\text{H}_{10}} \right) = 3.38 \text{ mol C}_4\text{H}_{10}$

$3.38 \text{ mol C}_4\text{H}_{10} \left(\dfrac{13 \text{ mol O}_2}{2 \text{ mol C}_4\text{H}_{10}} \right) \left(\dfrac{32.0 \text{ g O}_2}{1 \text{ mol O}_2} \right) = 703 \text{ g O}_2$

(c) $3.38 \text{ mol C}_4\text{H}_{10} \left(\dfrac{8 \text{ mol CO}_2}{2 \text{ mol C}_4\text{H}_{10}} \right) \left(\dfrac{44.0 \text{ g CO}_2}{1 \text{ mol CO}_2} \right) = 595 \text{ g CO}_2$

(d) $3.38 \text{ mol C}_4\text{H}_{10} \left(\dfrac{10 \text{ mol H}_2\text{O}}{2 \text{ mol C}_4\text{H}_{10}} \right) \left(\dfrac{18.0 \text{ g H}_2\text{O}}{1 \text{ mol H}_2\text{O}} \right) = 304 \text{ g H}_2\text{O}$

(e) $196 \text{ g C}_4\text{H}_{10} + 703 \text{ g O}_2 = 899 \text{ g reactants}$
$595 \text{ g CO}_2 + 304 \text{ g H}_2\text{O} = 899 \text{ g products}$

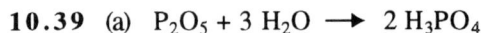

10.39 (a) $\text{P}_2\text{O}_5 + 3 \text{ H}_2\text{O} \longrightarrow 2 \text{ H}_3\text{PO}_4$

$75.0 \text{ g H}_3\text{PO}_4 \left(\dfrac{1 \text{ mol H}_3\text{PO}_4}{98.0 \text{ g H}_3\text{PO}_4} \right) = 0.765 \text{ mol H}_3\text{PO}_4$

$0.765 \text{ mol H}_3\text{PO}_4 \left(\dfrac{1 \text{ mol P}_2\text{O}_5}{2 \text{ mol H}_3\text{PO}_4} \right) \left(\dfrac{142.0 \text{ g P}_2\text{O}_5}{1 \text{ mol P}_2\text{O}_5} \right) = 54.3 \text{ g P}_2\text{O}_5$

(b) $0.765 \text{ mol H}_3\text{PO}_4 \left(\dfrac{3 \text{ mol H}_2\text{O}}{2 \text{ mol H}_3\text{PO}_4} \right) \left(\dfrac{18.0 \text{ g H}_2\text{O}}{1 \text{ mol H}_2\text{O}} \right) = 20.7 \text{ g H}_2\text{O}$

10.40 (a) $2.50 \text{ mol C}_6\text{H}_{12}\text{O}_6 \left(\dfrac{2 \text{ mol C}_2\text{H}_5\text{OH}}{1 \text{ mol C}_6\text{H}_{12}\text{O}_6} \right) \left(\dfrac{46.0 \text{ g C}_2\text{H}_5\text{OH}}{1 \text{ mol C}_2\text{H}_5\text{OH}} \right)$

$= 230 \text{ g C}_2\text{H}_5\text{OH} \quad (2.30 \times 10^2 \text{ g C}_2\text{H}_5\text{OH})$

(b) $2.50 \text{ mol C}_6\text{H}_{12}\text{O}_6 \left(\dfrac{2 \text{ mol CO}_2}{1 \text{ mol C}_6\text{H}_{12}\text{O}_6} \right) \left(\dfrac{44.0 \text{ g CO}_2}{1 \text{ mol CO}_2} \right)$

$= 220 \text{ g CO}_2 \quad (2.20 \times 10^2 \text{ g CO}_2)$

10.41 (a) $25.0 \text{ g CaCO}_3 \left(\dfrac{1 \text{ mol CaCO}_3}{100.1 \text{ g CaCO}_3} \right) \left(\dfrac{2 \text{ mol HC}_2\text{H}_3\text{O}_2}{1 \text{ mol CaCO}_3} \right) = 0.500 \text{ mol HC}_2\text{H}_3\text{O}_2$

(b) $75.0 \text{ g HC}_2\text{H}_3\text{O}_2 \left(\dfrac{1 \text{ mol HC}_2\text{H}_3\text{O}_2}{60.0 \text{ g HC}_2\text{H}_3\text{O}_2} \right) \left(\dfrac{1 \text{ mol CaCO}_3}{2 \text{ mol HC}_2\text{H}_3\text{O}_2} \right) = 0.625 \text{ mol CaCO}_3$

10.42 theoretical yield:

$15.7 \text{ g C}_7\text{H}_6\text{O}_3 \left(\dfrac{1 \text{ mol C}_7\text{H}_6\text{O}_3}{138.0 \text{ g C}_7\text{H}_6\text{O}_3} \right) \left(\dfrac{1 \text{ mol C}_8\text{H}_8\text{O}_3}{1 \text{ mol C}_7\text{H}_6\text{O}_3} \right) \left(\dfrac{152.0 \text{ g C}_8\text{H}_8\text{O}_3}{1 \text{ mol C}_8\text{H}_8\text{O}_3} \right)$

$= 17.3 \text{ g C}_8\text{H}_8\text{O}_3$

percentage yield: $\dfrac{10.4 \text{ g}}{17.3 \text{ g}} \times 100\% = 60.1\%$

10.43 theoretical yield:

$9.50 \text{ g MgCl}_2 \left(\dfrac{1 \text{ mol MgCl}_2}{95.3 \text{ g MgCl}_2} \right) \left(\dfrac{1 \text{ mol Mg(OH)}_2}{1 \text{ mol MgCl}_2} \right) \left(\dfrac{58.3 \text{ g Mg(OH)}_2}{1 \text{ mol Mg(OH)}_2} \right)$

$= 5.81 \text{ g Mg(OH)}_2$

percentage yield: $\dfrac{5.06 \text{ g}}{5.81 \text{ g}} \times 100\% = 87.1\%$

10.44 Theoretical yield:

$176 \text{ g HC}_2\text{H}_3\text{O}_2 \left(\dfrac{1 \text{ mol HC}_2\text{H}_3\text{O}_2}{60.0 \text{ g HC}_2\text{H}_3\text{O}_2} \right) \left(\dfrac{1 \text{ mol C}_4\text{H}_8\text{O}_2}{1 \text{ mol HC}_2\text{H}_3\text{O}_2} \right) \left(\dfrac{88.0 \text{ g C}_4\text{H}_8\text{O}_2}{1 \text{ mol C}_4\text{H}_8\text{O}_2} \right)$

$= 258 \text{ g C}_4\text{H}_8\text{O}_2$

Percentage yield: $\dfrac{215 \text{ g}}{258 \text{ g}} \times 100\% = 83.3\%$

10.45 (a) If you assume Al is the limiting reagent:

$5.75 \text{ g Al} \left(\dfrac{1 \text{ mol Al}}{27.0 \text{ g Al}} \right) \left(\dfrac{3 \text{ mol S}}{2 \text{ mol Al}} \right) \left(\dfrac{32.1 \text{ g S}}{1 \text{ mol S}} \right) = 10.3 \text{ g S}$

There is not enough S to react with 5.75 g Al, so S is the limiting reagent.

Alternatively, if you assume S is the limiting reagent:

$8.35 \text{ g S} \left(\dfrac{1 \text{ mol S}}{32.1 \text{ g S}} \right) \left(\dfrac{2 \text{ mol Al}}{3 \text{ mol S}} \right) \left(\dfrac{27.0 \text{ g Al}}{1 \text{ mol Al}} \right) = 4.68 \text{ g Al}$

(**10.45a** continues on next page.)

10.45 (continued):

There is more than enough Al to react with 8.35 g S, so S is the limiting reagent.

You must use the limiting reagent in the final calculation:

(b) $8.35 \text{ g S}\left(\dfrac{1 \text{ mol S}}{32.1 \text{ g S}}\right)\left(\dfrac{1 \text{ mol Al}_2\text{S}_3}{3 \text{ mol S}}\right)\left(\dfrac{150.3 \text{ g Al}_2\text{S}_3}{1 \text{ mol Al}_2\text{S}_3}\right) = 13.0 \text{ g Al}_2\text{S}_3$

10.46 If you assume Na is the limiting reagent:

$13.5 \text{ g Na}\left(\dfrac{1 \text{ mol Na}}{23.0 \text{ g Na}}\right)\left(\dfrac{2 \text{ mol H}_2\text{O}}{2 \text{ mol Na}}\right)\left(\dfrac{18.0 \text{ g H}_2\text{O}}{1 \text{ mol H}_2\text{O}}\right) = 10.6 \text{ g H}_2\text{O}$

There is more than ehough water to consume all of the sodium, so sodium is the limiting reagent.

If you assume water is the limiting reagent:

$12.8 \text{ g H}_2\text{O}\left(\dfrac{1 \text{ mol H}_2\text{O}}{18.0 \text{ g H}_2\text{O}}\right)\left(\dfrac{2 \text{ mol Na}}{2 \text{ mol H}_2\text{O}}\right)\left(\dfrac{23.0 \text{ g Na}}{1 \text{ mol Na}}\right) = 16.4 \text{ g Na}$

There is not enough sodium to consume all of the water, so sodium is the limiting reagent.

$13.5 \text{ g Na}\left(\dfrac{1 \text{ mol Na}}{23.0 \text{ g Na}}\right)\left(\dfrac{1 \text{ mol H}_2}{2 \text{ mol Na}}\right)\left(\dfrac{2.0 \text{ g H}_2}{1 \text{ mol H}_2}\right) = 0.59 \text{ g H}_2$

10.47 We will assume that H_2SO_4 is the limiting reagent and see if there is enough $NaHCO_3$ to consume all of it:

$125 \text{ g H}_2\text{SO}_4\left(\dfrac{1 \text{ mol H}_2\text{SO}_4}{98.1 \text{ g H}_2\text{SO}_4}\right)\left(\dfrac{2 \text{ mol NaHCO}_3}{1 \text{ mol H}_2\text{SO}_4}\right)\left(\dfrac{84.0 \text{ g NaHCO}_3}{1 \text{ mol NaHCO}_3}\right) = 214 \text{ g NaHCO}_3$

There is more than enough $NaHCO_3$ to react with 125 g of the H_2SO_4, so H_2SO_4 *is* the limiting reagent.

$125 \text{ g H}_2\text{SO}_4\left(\dfrac{1 \text{ mol H}_2\text{SO}_4}{98.1 \text{ g H}_2\text{SO}_4}\right)\left(\dfrac{2 \text{ mol CO}_2}{1 \text{ mol H}_2\text{SO}_4}\right)\left(\dfrac{44.0 \text{ g CO}_2}{1 \text{ mol CO}_2}\right) = 112 \text{ g CO}_2$

10.48 (a) An exothermic reaction liberates heat.

$CH_4 + 2 O_2 \longrightarrow CO_2 + 2 H_2O$

The sign of ΔH is negative.

(b) An endothermic reaction absorbs heat.

$6 CO_2 + 6 H_2O \longrightarrow C_6H_{12}O_6 + 6 H_2O$

The sign of ΔH is positive.

10.49 $400 \text{ g C}_3\text{H}_8\left(\dfrac{1 \text{ mol C}_3\text{H}_8}{44.0 \text{ g C}_3\text{H}_8}\right)\left(\dfrac{2220 \text{ kJ}}{1 \text{ mol C}_3\text{H}_8}\right) = 2.02 \times 10^4 \text{ kJ}$

10.50 $12.5 \text{ g CaCO}_3 \left(\dfrac{1 \text{ mol CaCO}_3}{100.1 \text{ g CaCO}_3} \right) \left(\dfrac{178 \text{ kJ}}{1 \text{ mol CaCO}_3} \right) = 22.2 \text{ kJ}$

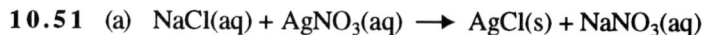

10.51 (a) $\text{NaCl(aq)} + \text{AgNO}_3\text{(aq)} \longrightarrow \text{AgCl(s)} + \text{NaNO}_3\text{(aq)}$

(b) If you assume NaCl is the limiting reagent:

$10.0 \text{ g NaCl} \left(\dfrac{1 \text{ mol NaCl}}{58.5 \text{ g NaCl}} \right) \left(\dfrac{1 \text{ mol AgNO}_3}{1 \text{ mol NaCl}} \right) \left(\dfrac{169.9 \text{ g AgNO}_3}{1 \text{ mol AgNO}_3} \right) = 29.0 \text{ g AgNO}_3$

There is not enough AgNO_3 to react with all of the NaCl, so AgNO_3 is the limiting reagent.

(c) Theoretical yield:

$10.0 \text{ g AgNO}_3 \left(\dfrac{1 \text{ mol AgNO}_3}{169.9 \text{ g AgNO}_3} \right) \left(\dfrac{1 \text{ mol AgCl}}{1 \text{ mol AgNO}_3} \right) \left(\dfrac{143.4 \text{ g AgCl}}{1 \text{ mol AgCl}} \right) = 8.44 \text{ g AgCl}$

(d) Percentage yield: $\dfrac{7.86 \text{ g}}{8.44 \text{ g}} \times 100\% = 93.1\%$

10.52 (a) $\text{Pb(NO}_3)_2\text{(aq)} + 2 \text{ KI(aq)} \longrightarrow \text{PbI}_2\text{(s)} + 2 \text{ KNO}_3\text{(aq)}$

(b) If you assume that $\text{Pb(NO}_3)_2$ is the limiting reagent:

$8.50 \text{ g Pb(NO}_3)_2 \left(\dfrac{1 \text{ mol Pb(NO}_3)_2}{331.2 \text{ g Pb(NO}_3)_2} \right) \left(\dfrac{2 \text{ mol KI}}{1 \text{ mol Pb(NO}_3)_2} \right) \left(\dfrac{166.0 \text{ g KI}}{1 \text{ mol KI}} \right) = 8.52 \text{ g KI}$

There is not enough KI present, so KI is the limiting reagent.

(c) $6.50 \text{ g KI} \left(\dfrac{1 \text{ mol KI}}{166.0 \text{ g KI}} \right) \left(\dfrac{1 \text{ mol PbI}_2}{2 \text{ mol KI}} \right) \left(\dfrac{461.0 \text{ g PbI}_2}{1 \text{ mol PbI}_2} \right) = 9.03 \text{ g PbI}_2$

(d) Percentage yield: $\dfrac{7.25 \text{ g}}{9.03 \text{ g}} \times 100\% = 80.3\%$

10.53 One approach to this problem is to begin by calculating the quantities of reactants required to produce 26.0 g of ispentyl acetate, and then make a correction for the 55% yield.

$26.0 \text{ g } \text{C}_7\text{H}_{14}\text{O}_2 \left(\dfrac{1 \text{ mol C}_7\text{H}_{14}\text{O}_2}{130 \text{ g C}_7\text{H}_{14}\text{O}_2} \right) \left(\dfrac{1 \text{ mol C}_2\text{H}_4\text{O}_2}{1 \text{ mol C}_7\text{H}_{14}\text{O}_2} \right) \left(\dfrac{60.0 \text{ g C}_2\text{H}_4\text{O}_2}{1 \text{ mol C}_2\text{H}_4\text{O}_2} \right)$

$= 12.0 \text{ g C}_2\text{H}_4\text{O}_2$

$26.0 \text{ g } \text{C}_7\text{H}_{14}\text{O}_2 \left(\dfrac{1 \text{ mol C}_7\text{H}_{14}\text{O}_2}{130 \text{ g C}_7\text{H}_{14}\text{O}_2} \right) \left(\dfrac{1 \text{ mol C}_5\text{H}_{12}\text{O}}{1 \text{ mol C}_7\text{H}_{14}\text{O}_2} \right) \left(\dfrac{88.0 \text{ g C}_5\text{H}_{12}\text{O}}{1 \text{ mol C}_5\text{H}_{12}\text{O}} \right)$

$= 17.6 \text{ g C}_5\text{H}_{12}\text{O}$

(**10.53** continues on next page.)

10.53 (continued):

If the reaction gave a 100% yield, 12.0 g of acetic acid and 17.6 g of isopentyl alcohol would be used. However, the reaction proceeds in 55% yield. To correct for the lower yield, divide each quantity by 0.55. This gives the quantity that each of the calculated values is 55% of:

$$\frac{12.0 \text{ g C}_2\text{H}_4\text{O}_2}{0.55} = 22 \text{ g C}_2\text{H}_4\text{O}_2 \qquad \frac{17.6 \text{ g C}_5\text{H}_{12}\text{O}}{0.55} = 32 \text{ g C}_5\text{H}_{12}\text{O}$$

10.54 (a) energy absorbed: 436 kJ + 243 kJ = 679 kJ

energy liberated: $2 \times 432 = 864$ kJ

net energy liberated: $\Delta H = +679$ kJ - 864 kJ = -185 kJ

(b) energy absorbed: 2×366 kJ = 732 kJ

energy liberated: 436 kJ + 193 kJ = 629 kJ

net energy absorbed: $\Delta H = +732$ kJ - 629 kJ = +103 kJ

10.55 (c) and (e) are not possible for the following reasons:

(c) The molecular mass of HgO is greater than that of Hg, so the number of grams of Hg that can be produced from 10 g of HgO must be less than 10 g.

(e) The molecular mass of C_2H_4 is less than that of C_2H_5OH, so it is impossible to produce 10 g of C_2H_4 from 10 g of C_2H_5OH.

10.56 The hydrogen to carbon ratio is greater in ethane (6:2, which is the same as 18:6) than it is in hexane (14:6). Thus, there is more hydrogen in an equal mass of ethane, and this will lead to a greater mass of water.

10.57 The carbon to hydrogen ratio in propane is 3:8 or 12:32. The carbon to hydrogen ratio in butane is 4:10 or 12:30. Thus, there is a greater percentage of carbon in butane (fewer hydrogens for the same number of carbons). If both produce carbon dioxide at the same rate, propane (the compound with a lower percentage of carbon) will run out of carbon first.

CHAPTER 11

The Gas Laws

11.1 The word *kinetic* refers to motion. Gas molecules are in constant motion, colliding with one another and with other objects in their paths.

11.2 P_{gas} = 764 torr + 7 torr = 771 torr

11.3 (a) The pressure experienced underwater comes from the weight of the water above. Imagine that a column of water (much like the column of mercury in a barometer) exists between the surface of the pool and the level of the ear. The weight of water in this column exerts a force per unit area on the ear.
(b) The greater the depth, the taller the column of water above the ear, and hence, the greater the pressure.

11.4 (a) $30.0 \text{ mL} \left(\dfrac{750.0 \text{ torr}}{600.0 \text{ torr}} \right) = 37.5 \text{ mL}$ (b) $500.0 \text{ torr} \left(\dfrac{1.50 \text{ L}}{1.00 \text{ L}} \right) = 750 \text{ torr}$

(c) $45.0 \text{ mL} \left(\dfrac{1.00 \text{ atm}}{1.00 \text{ atm}} \right) = 45.0 \text{ mL}$ (d) $3.60 \text{ L} \left(\dfrac{1.20 \text{ atm}}{0.800 \text{ atm}} \right) = 5.40 \text{ L}$

(e) $700.0 \text{ torr} \left(\dfrac{0.0840 \text{ L}}{5.20 \text{ L}} \right) = 11.3 \text{ torr}$ (f) $76.0 \text{ mL} \left(\dfrac{532 \text{ torr}}{480 \text{ torr}} \right) = 84.2 \text{ mL}$

(g) $146 \text{ torr} \left(\dfrac{18.2 \text{ L}}{11.4 \text{ L}} \right) = 233 \text{ torr}$

11.5 (a) 300 K (b) 233 K (c) 546 K (d) – 116°C (e) 142°C (f) – 273°C

11.6 (a) $205 \text{ torr} \left(\dfrac{636 \text{ K}}{212 \text{ K}} \right) = 615 \text{ torr}$ (b) $1.40 \text{ atm} \left(\dfrac{216 \text{ K}}{324 \text{ K}} \right) = 0.933 \text{ atm}$

(c) $575 \text{ torr} \left(\dfrac{400 \text{ K}}{300 \text{ K}} \right) = 767 \text{ torr}$ (d) $330 \text{ K} \left(\dfrac{3.00 \text{ atm}}{2.00 \text{ atm}} \right) = 495 \text{ K} = 222°C$

(e) $298 \text{ K} \left(\dfrac{1.00 \text{ atm}}{0.873 \text{ atm}} \right) = 341 \text{ K} = 68°C$ (f) $645 \text{ torr} \left(\dfrac{273 \text{ K}}{373 \text{ K}} \right) = 472 \text{ torr}$

(g) $195 \text{ K} \left(\dfrac{760 \text{ torr}}{555 \text{ torr}} \right) = 267 \text{ K} = - 6°C$

11.7 (a) $50.0 \text{ mL}\left(\dfrac{400 \text{ K}}{300 \text{ K}}\right) = 66.7 \text{ mL}$ (b) $3.50 \text{ mL}\left(\dfrac{273 \text{ K}}{373 \text{ K}}\right) = 2.56 \text{ mL}$

   (c) $200 \text{ K}\left(\dfrac{0.700 \text{ L}}{0.400 \text{ L}}\right) = 350 \text{ K}$ (d) $298 \text{ K}\left(\dfrac{1020 \text{ mL}}{68.0 \text{ mL}}\right) = 4470 \text{ K}$ $(4.47 \times 10^3 \text{ K})$

   (e) $0.250 \text{ L}\left(\dfrac{546 \text{ K}}{273 \text{ K}}\right) = 0.500 \text{ L}$

11.8 (a) $75.0 \text{ mL}\left(\dfrac{732 \text{ torr}}{946 \text{ torr}}\right)\left(\dfrac{273 \text{ K}}{298 \text{ K}}\right) = 53.2 \text{ mL}$

   (b) $310 \text{ K}\left(\dfrac{60.0 \text{ mL}}{55.0 \text{ mL}}\right)\left(\dfrac{715 \text{ torr}}{785 \text{ torr}}\right) = 308 \text{ K} = 35°\text{C}$

   (c) $1.00 \text{ atm}\left(\dfrac{4.20 \text{ L}}{5.00 \text{ L}}\right)\left(\dfrac{200 \text{ K}}{415 \text{ K}}\right) = 0.405 \text{ atm}$

   (d) $17.5 \text{ L}\left(\dfrac{1.40 \text{ atm}}{0.916 \text{ atm}}\right)\left(\dfrac{273 \text{ K}}{384 \text{ K}}\right) = 19.0 \text{ L}$

   (e) $195 \text{ K}\left(\dfrac{0.900 \text{ L}}{0.0450 \text{ L}}\right)\left(\dfrac{2.50 \text{ atm}}{0.283 \text{ atm}}\right) = 34{,}500 \text{ K}$ $(3.45 \times 10^4 \text{ K})$

   (f) $463 \text{ torr}\left(\dfrac{0.355 \text{ L}}{0.500 \text{ L}}\right)\left(\dfrac{173 \text{ K}}{546 \text{ K}}\right) = 104 \text{ torr}$

11.9 $0.400 \text{ L}\left(\dfrac{1.23 \text{ atm}}{1.00 \text{ atm}}\right)\left(\dfrac{273 \text{ K}}{330 \text{ K}}\right) = 0.407 \text{ L}$

11.10 $72.0 \text{ mL}\left(\dfrac{695 \text{ torr}}{760 \text{ torr}}\right)\left(\dfrac{273 \text{ K}}{400 \text{ K}}\right) = 44.9 \text{ mL}$

11.11 $1.00 \text{ atm}\left(\dfrac{6.00 \text{ L}}{0.455 \text{ L}}\right)\left(\dfrac{263 \text{ K}}{350 \text{ K}}\right) = 9.91 \text{ atm}$

11.12 $295 \text{ K}\left(\dfrac{50.0 \text{ mL}}{40.0 \text{ mL}}\right)\left(\dfrac{760 \text{ torr}}{666 \text{ torr}}\right) = 421 \text{ K} = 148°\text{C}$

11.13 (a) $5.60 \text{ L}\left(\dfrac{1 \text{ mol}}{22.4 \text{ L}}\right) = 0.250 \text{ mol}$

(b) $V_{STP} = 3.20 \text{ L}\left(\dfrac{273 \text{ K}}{298 \text{ K}}\right) = 2.93 \text{ L}$

$2.93 \text{ L}\left(\dfrac{1 \text{ mol}}{22.4 \text{ L}}\right) = 0.131 \text{ mol}$

(c) $V_{STP} = 45.0 \text{ mL}\left(\dfrac{745 \text{ torr}}{760 \text{ torr}}\right)\left(\dfrac{273 \text{ K}}{228 \text{ K}}\right) = 52.8 \text{ mL} = 0.0528 \text{ L}$

$0.0528 \text{ L}\left(\dfrac{1 \text{ mol}}{22.4 \text{ L}}\right) = 0.00236 \text{ mol}$

(d) $V_{STP} = 5.00 \text{ L}\left(\dfrac{855 \text{ torr}}{760 \text{ torr}}\right)\left(\dfrac{273 \text{ K}}{516 \text{ K}}\right) = 2.98 \text{ L}$

$2.98 \text{ L}\left(\dfrac{1 \text{ mol}}{22.4 \text{ L}}\right) = 0.133 \text{ mol}$

(e) $V_{STP} = 18.0 \text{ mL}\left(\dfrac{17.6 \text{ atm}}{1.00 \text{ atm}}\right)\left(\dfrac{273 \text{ K}}{90 \text{ K}}\right) = 960 \text{ mL} = 0.96 \text{ L}$

$0.96 \text{ L}\left(\dfrac{1 \text{ mol}}{22.4 \text{ L}}\right) = 0.043 \text{ mol}$

11.14 $V_{STP} = 1.00 \text{ mL}\left(\dfrac{1.00 \text{ torr}}{760 \text{ torr}}\right)\left(\dfrac{273 \text{ K}}{1000 \text{ K}}\right) = 3.59 \times 10^{-4} \text{ mL}$

$3.59 \times 10^{-4} \text{ mL}\left(\dfrac{1 \text{ L}}{1000 \text{ mL}}\right)\left(\dfrac{1 \text{ mol}}{22.4 \text{ L}}\right)\left(\dfrac{6.02 \times 10^{23} \text{ molecules}}{1 \text{ mol}}\right)$

$= 9.65 \times 10^{15} \text{ molecules}$

11.15 (a) $\dfrac{0.0821 \text{ L} \cdot \text{atm}}{\text{mol} \cdot \text{K}}\left(\dfrac{1000 \text{ mL}}{1 \text{ L}}\right) = \dfrac{82.1 \text{ mL} \cdot \text{atm}}{\text{mol} \cdot \text{K}}$

(b) $\dfrac{0.0821 \text{ L} \cdot \text{atm}}{\text{mol} \cdot \text{K}}\left(\dfrac{760 \text{ torr}}{1 \text{ atm}}\right) = \dfrac{62.4 \text{ L} \cdot \text{torr}}{\text{mol} \cdot \text{K}}$

11.16 $(745 \text{ torr})(V) = 0.275 \text{ mol} \left(\dfrac{62.4 \text{ L} \cdot \text{torr}}{\text{mol} \cdot \text{K}} \right)(195 \text{ K})$

$V = \dfrac{(0.275)(62.4)(195)\text{L}}{(745)} = 4.49 \text{ L}$

11.17 $(P)(1.00 \text{ L}) = 2.00 \text{ mol} \left(\dfrac{0.0821 \text{ L} \cdot \text{atm}}{\text{mol} \cdot \text{K}} \right)(295 \text{ K})$

$P = \dfrac{(2.00)(0.0821)(295)\text{atm}}{(1.00)} = 48.4 \text{ atm}$

11.18 $0.640 \text{ g O}_2 \left(\dfrac{1 \text{ mol O}_2}{32.0 \text{ g O}_2} \right) = 0.0200 \text{ mol O}_2$

$(747 \text{ torr})(575 \text{ mL}) = 0.0200 \text{ mol} \left(\dfrac{6.24 \times 10^4 \text{ mL} \cdot \text{torr}}{\text{mol} \cdot \text{K}} \right)(T)$

$\dfrac{(747)(575)\text{K}}{(0.0200)(6.24 \times 10^4)} = T = 344 \text{ K} = 71°C$

11.19 $(1.15 \text{ atm})(276 \text{ mL}) = (n) \left(\dfrac{82.1 \text{ mL} \cdot \text{atm}}{\text{mol} \cdot \text{K}} \right)(249 \text{ K})$

$\dfrac{(1.15)(276)\text{mol}}{(82.1)(249)} = n = 0.0155 \text{ mol}$

11.20 $6.14 \text{ L C}_2\text{H}_6 \left(\dfrac{4 \text{ L CO}_2}{2 \text{ L C}_2\text{H}_6} \right) = 12.3 \text{ L CO}_2$

11.21 $6.14 \text{ L C}_2\text{H}_6 \left(\dfrac{7 \text{ L O}_2}{2 \text{ L C}_2\text{H}_6} \right) = 21.5 \text{ L O}_2$

11.22 $4.50 \text{ g Ag}_2\text{O} \left(\dfrac{1 \text{ mol Ag}_2\text{O}}{231.8 \text{ g Ag}_2\text{O}} \right)\left(\dfrac{1 \text{ mol O}_2}{2 \text{ mol Ag}_2\text{O}} \right)\left(\dfrac{22.4 \text{ L O}_2}{1 \text{ mol O}_2} \right) = 0.217 \text{ L O}_2$

11.23 $1.75 \text{ L C}_4\text{H}_{10}\left(\dfrac{8 \text{ L CO}_2}{2 \text{ L C}_4\text{H}_{10}}\right) = 7.00 \text{ L CO}_2$ (at 298 K, 765 torr)

$7.00 \text{ L}\left(\dfrac{528 \text{ K}}{298 \text{ K}}\right)\left(\dfrac{765 \text{ torr}}{395 \text{ torr}}\right) = 24.0 \text{ L}$

11.24 $V_{STP} = 1.35 \text{ L}\left(\dfrac{273 \text{ K}}{458 \text{ K}}\right)\left(\dfrac{635 \text{ torr}}{760 \text{ torr}}\right) = 0.672 \text{ L}$

$0.672 \text{ L}\left(\dfrac{1 \text{ mol O}_2}{22.4 \text{ L}}\right) = 0.0300 \text{ mol O}_2$

$0.0300 \text{ mol O}_2\left(\dfrac{1 \text{ mol S}}{1 \text{ mol O}_2}\right)\left(\dfrac{32.1 \text{ g S}}{1 \text{ mol S}}\right) = 0.963 \text{ g S}$

11.25 $5.00 \text{ g H}_2\text{O}_2\left(\dfrac{1 \text{ mol H}_2\text{O}_2}{34.0 \text{ g H}_2\text{O}_2}\right)\left(\dfrac{1 \text{ mol O}_2}{2 \text{ mol H}_2\text{O}_2}\right) = 0.0735 \text{ mol O}_2$

$0.0735 \text{ mol O}_2\left(\dfrac{22.4 \text{ L O}_2}{1 \text{ mol O}_2}\right) = 1.65 \text{ L O}_2$ (at STP)

$V_f = 1.65 \text{ L}\left(\dfrac{296 \text{ K}}{273 \text{ K}}\right)\left(\dfrac{760 \text{ torr}}{712 \text{ torr}}\right) = 1.91 \text{ L}$

11.26 $V_{STP} = 6.00 \text{ L}\left(\dfrac{732 \text{ torr}}{760 \text{ torr}}\right)\left(\dfrac{273 \text{ K}}{373 \text{ K}}\right) = 4.23 \text{ L}$

$4.23 \text{ L}\left(\dfrac{1 \text{ mol}}{22.4 \text{ L}}\right) = 0.189 \text{ mol}$

$\text{molar mass} = \dfrac{16.0 \text{ g}}{0.189 \text{ mol}} = 84.7 \text{ g/mol}$

11.27 $V_{STP} = 1.50 \text{ L}\left(\dfrac{777 \text{ torr}}{760 \text{ torr}}\right)\left(\dfrac{273 \text{ K}}{218 \text{ K}}\right) = 1.92 \text{ L}$

$1.92 \text{ L}\left(\dfrac{1 \text{ mol}}{22.4 \text{ L}}\right) = 0.0857 \text{ mol}$

$\text{molar mass} = \dfrac{2.40 \text{ g}}{0.0857 \text{ mol}} = 28.0 \text{ g/mol}$

11.28 $V_{STP} = 0.500 \text{ L} \left(\dfrac{0.900 \text{ atm}}{1.00 \text{ atm}} \right) \left(\dfrac{273 \text{ K}}{283 \text{ K}} \right) = 0.434 \text{ L}$

$0.434 \text{ L} \left(\dfrac{1 \text{ mol}}{22.4 \text{ L}} \right) = 0.0194 \text{ mol}$

$\text{molar mass} = \dfrac{2.67 \text{ g}}{0.0194 \text{ mol}} = 138 \text{ g/mol}$

11.29 135 torr + 624 torr + 43 torr + 13 torr = 815 torr

11.30 total number of moles = 0.25 mol + 0.33 mol + 0.42 mol = 1.00 mol

$P_{He} = \dfrac{0.25}{1.00}(2.00 \text{ atm}) = 0.50 \text{ atm}$ $\qquad$ $P_{Ar} = \dfrac{0.33}{1.00}(2.00 \text{ atm}) = 0.66 \text{ atm}$

$P_{Ne} = \dfrac{0.42}{1.00}(2.00 \text{ atm}) = 0.84 \text{ atm}$

11.31 total number of moles = 1.00 mol + 2.00 mol + 7.00 mol = 10.00 mol

$P_{O_2} = \dfrac{1.00}{10.00}(1.20 \text{ atm}) = 0.120 \text{ atm}$ $\qquad$ $P_{H_2} = \dfrac{2.00}{10.00}(1.20 \text{ atm}) = 0.240 \text{ atm}$

$P_{N_2} = \dfrac{7.00}{10.00}(1.20 \text{ atm}) = 0.840 \text{ atm}$

11.32 $2.80 \text{ g N}_2 \left(\dfrac{1 \text{ mol N}_2}{28.0 \text{ g N}_2} \right) = 0.100 \text{ mol N}_2$ $\quad$ $6.40 \text{ g O}_2 \left(\dfrac{1 \text{ mol O}_2}{32.0 \text{ g O}_2} \right) = 0.200 \text{ mol O}_2$

total moles = 0.100 mol + 0.200 mol = 0.300 mol

$P_{N_2} = \dfrac{0.100}{0.300}(720 \text{ torr}) = 240 \text{ torr}$ $\qquad$ $P_{O_2} = \dfrac{0.200}{0.300}(720 \text{ torr}) = 480 \text{ torr}$

11.33 $P_{H_2} = 770.0 \text{ torr} - 21.1 \text{ torr} = 748.9 \text{ torr}$

11.34 $P_{N_2} = 758.8 \text{ torr} - 19.8 \text{ torr} = 739.0 \text{ torr}$

$0.250 \text{ L} \left(\dfrac{739.0 \text{ torr}}{760.0 \text{ torr}} \right) \left(\dfrac{273 \text{ K}}{295 \text{ K}} \right) = 0.225 \text{ L}$ $\quad$ $0.225 \text{ L} \left(\dfrac{1 \text{ mol}}{22.4 \text{ L}} \right) = 0.0100 \text{ mol}$

11.35 (a) Gas molecules are in constant motion, colliding with one another and with objects in their paths.
(b) The intermolecular forces of attraction are negligible.
(c) The velocity of gaseous molecules increases with increasing temperature.
(d) An elastic collision is one that occurs with no loss of energy.
(e) At temperatures near the boiling point, intermolecular forces of attraction are not negligible. Under high pressure, the volumes of molecules are not negligible.

11.36 (a) Pressure is the force per unit area.

(b) Collisions of gas molecules on a surface exert a force against the area they collide with.

(c) The pressure exerted by gas molecules on the mercury pool outside of the glass tube equals the pressure created by the column of mercury in the tube. The column of mercury inside the tube rises and falls to match the corresponding atmospheric pressure.

(d) If one arm of a mercury U-tube is connected to a vessel containing a gas and the other arm is open to the atmosphere, the difference in heights of the mercury in the two arms will equal the difference between the pressure of the gas and the atmospheric pressure. If a barometer is read to determine the atmospheric pressure, the height difference in the arms of the U-tube may be added or subtracted from the barometric pressure (depending upon which has the greater pressure) to give the pressure of the gas.

(e) A vacuum is an absence of gaseous molecules.

11.37 $V_f = 43.0 \text{ mL} \left(\dfrac{726 \text{ torr}}{942 \text{ torr}} \right) = 33.1 \text{ mL}$

11.38 $V_f = 1.50 \text{ L} \left(\dfrac{12.0 \text{ atm}}{0.985 \text{ atm}} \right) = 18.3 \text{ L}$

11.39 $P_f = 755 \text{ torr} \left(\dfrac{9.80 \text{ L}}{4.40 \text{ L}} \right) = 1680 \text{ torr} \quad (1.68 \times 10^3 \text{ torr})$

11.40 $P_f = 780 \text{ torr} \left(\dfrac{0.735 \text{ L}}{1.00 \text{ L}} \right) = 573 \text{ torr}$

11.41 At absolute zero, the kinetic energy is zero as are all molecular velocities.

11.42 $P_f = 1.00 \text{ atm} \left(\dfrac{373 \text{ K}}{298 \text{ K}} \right) = 1.25 \text{ atm}$

11.43 $T_f = 310 \text{ K} \left(\dfrac{600 \text{ torr}}{760 \text{ torr}} \right) = 245 \text{ K} = -28°\text{C}$

11.44 $V_f = 755 \text{ mL} \left(\dfrac{309 \text{ K}}{293 \text{ K}} \right) = 796 \text{ mL}$

11.45 $V_f = 2.65 \text{ L} \left(\dfrac{195 \text{ K}}{295 \text{ K}} \right) = 1.75 \text{ L}$

11.46 Standard temperature: $0°C = 273 \text{ K}$
Standard pressure: $1.00 \text{ atm} = 760.0 \text{ torr} = 1.013 \times 10^5 \text{ Pa}$

11.47 $0.105 \text{ atm} = 79.8 \text{ torr}$
$V_f = 4.20 \text{ L} \left(\dfrac{243 \text{ K}}{300 \text{ K}} \right) \left(\dfrac{740 \text{ torr}}{79.8 \text{ torr}} \right) = 31.5 \text{ L}$

11.48 $T_f = 308 \text{ K} \left(\dfrac{0.550 \text{ L}}{0.450 \text{ L}} \right) \left(\dfrac{1.15 \text{ atm}}{1.25 \text{ atm}} \right) = 346 \text{ K} = 73°C$

11.49 $V_f = 575 \text{ mL} \left(\dfrac{1.05 \text{ atm}}{1.55 \text{ atm}} \right) \left(\dfrac{288 \text{ K}}{298 \text{ K}} \right) = 376 \text{ mL}$

11.50 $P_f = 535 \text{ torr} \left(\dfrac{75.0 \text{ mL}}{55.0 \text{ mL}} \right) \left(\dfrac{228 \text{ K}}{301 \text{ K}} \right) = 553 \text{ torr}$

11.51 $V_f = 22.4 \text{L} \left(\dfrac{298 \text{ K}}{273 \text{ K}} \right) \left(\dfrac{1.00 \text{ atm}}{1.00 \text{ atm}} \right) = 24.5 \text{ L}$

11.52 $25.0 \text{ mL} \left(\dfrac{1 \text{ L}}{1000 \text{ mL}} \right) \left(\dfrac{1 \text{ mol}}{22.4 \text{ L}} \right) = 0.00112 \text{ mol}$

11.53 $V_{STP} = 12.4 \text{ L} \left(\dfrac{273 \text{ K}}{310 \text{ K}} \right) \left(\dfrac{680 \text{ torr}}{760 \text{ torr}} \right) = 9.77 \text{ L}$

$9.77 \text{ L} \left(\dfrac{1 \text{ mol}}{22.4 \text{ L}} \right) = 0.436 \text{ mol}$

11.54 $V_{STP} = 125 \text{ mL} \left(\dfrac{273 \text{ K}}{300 \text{ K}} \right) \left(\dfrac{735 \text{ torr}}{760 \text{ torr}} \right) = 110 \text{ mL} \quad (1.10 \times 10^2 \text{ mL})$

$110 \text{ mL} \left(\dfrac{1 \text{ L}}{1000 \text{ mL}} \right) \left(\dfrac{1 \text{ mol}}{22.4 \text{ L}} \right) = 0.00491 \text{ mol}$

11.55 $V_f = 22.4 \text{ L} \left(\dfrac{293 \text{ K}}{273 \text{ K}} \right) \left(\dfrac{101.3 \text{ kPa}}{100.0 \text{ kPa}} \right) = 24.4 \text{ L}$

11.56 $(P)(3.00 \text{ L}) = (0.500 \text{ mol}) \left(\dfrac{0.0821 \text{ L} \cdot \text{atm}}{\text{mol} \cdot \text{K}} \right) (348 \text{ K})$

$P = \dfrac{(0.500)(0.0821)(348) \text{atm}}{(3.00)} = 4.76 \text{ atm} \quad (= 3.62 \times 10^3 \text{ torr})$

11.57 $15.0 \text{ g SO}_2 \left(\dfrac{1 \text{ mol SO}_2}{64.1 \text{ g SO}_2} \right) = 0.234 \text{ mol SO}_2$

$(P)(5.00 \text{ L}) = (0.234 \text{ mol}) \left(\dfrac{0.0821 \text{ L} \cdot \text{atm}}{\text{mol} \cdot \text{K}} \right) (302 \text{ K})$

$P = \dfrac{(0.234)(0.0821)(302) \text{atm}}{(5.00)} = 1.16 \text{ atm} \quad (= 882 \text{ torr})$

11.58 $(744 \text{ torr})(325 \text{ mL}) = (0.0123 \text{ mol}) \left(\dfrac{6.24 \times 10^4 \text{ mL} \cdot \text{torr}}{\text{mol} \cdot \text{K}} \right) (T)$

$\dfrac{(744)(325) \text{K}}{(0.0123)(6.24 \times 10^4)} = T = 315 \text{ K} = 42°\text{C}$

11.59 $(775 \text{ torr})(1.75 \text{ L}) = (n) \left(\dfrac{62.4 \text{ L} \cdot \text{torr}}{\text{mol} \cdot \text{K}} \right) (291 \text{ K})$

$\dfrac{(775)(1.75) \text{mol}}{(62.4)(291)} = n = 0.0747 \text{ mol}$

11.60 (a) $1.55 \text{ L } C_4H_8 \left(\dfrac{4 \text{ L } CO_2}{1 \text{ L } C_4H_8} \right) = 6.20 \text{ L } CO_2$

   (b) $1.55 \text{ L } C_4H_8 \left(\dfrac{6 \text{ L } O_2}{1 \text{ L } C_4H_8} \right) = 9.30 \text{ L } O_2$

11.61 $13.4 \text{ L } NH_3 \left(\dfrac{3 \text{ L } H_2}{2 \text{ L } NH_3} \right) = 20.1 \text{ L } H_2$

11.62 $576 \text{ ml } CO_2 \left(\dfrac{1 \text{ L } CO_2}{1000 \text{ mL } CO_2} \right) \left(\dfrac{1 \text{ mol } CO_2}{22.4 \text{ L } CO_2} \right) = 0.0257 \text{ mol } CO_2$

   $0.0257 \text{ mol } CO_2 \left(\dfrac{1 \text{ mol } BaCO_3}{1 \text{ mol } CO_2} \right) \left(\dfrac{197.3 \text{ g } BaCO_3}{1 \text{ mol } BaCO_3} \right) = 5.07 \text{ g } BaCO_3$

11.63 $11.6 \text{ g } Al \left(\dfrac{1 \text{ mol } Al}{27.0 \text{ g } Al} \right) \left(\dfrac{3 \text{ mol } H_2}{2 \text{ mol } Al} \right) \left(\dfrac{22.4 \text{ L } H_2}{1 \text{ mol } H_2} \right) = 14.4 \text{ L } H_2$

11.64 $135 \text{ mL } N_2H_4 \left(\dfrac{2 \text{ mL } NO_2}{1 \text{ ml } N_2H_4} \right) \left(\dfrac{598 \text{ K}}{428 \text{ K}} \right) \left(\dfrac{646 \text{ torr}}{525 \text{ torr}} \right) = 464 \text{ mL } NO_2$

11.65 $1.00 \text{ g } KClO_3 \left(\dfrac{1 \text{ mol } KClO_3}{122.6 \text{ g } KClO_3} \right) \left(\dfrac{3 \text{ mol } O_2}{2 \text{ mol } KClO_3} \right) \left(\dfrac{22.4 \text{ L } O_2}{1 \text{ mol } O_2} \right) \left(\dfrac{1000 \text{ mL } O_2}{1 \text{ L } O_2} \right)$

   $= 274 \text{ mL } O_2 \text{ (at STP)}$

   $274 \text{ mL} \left(\dfrac{293 \text{ K}}{273 \text{ K}} \right) \left(\dfrac{760 \text{ torr}}{768 \text{ torr}} \right) = 291 \text{ mL} \text{ (at } 20°C, 768 \text{ torr)}$

11.66 $175 \text{ mL } NO \left(\dfrac{273 \text{ K}}{308 \text{ K}} \right) \left(\dfrac{745 \text{ torr}}{760 \text{ torr}} \right) = 152 \text{ mL } NO \text{ (at STP)}$

   $152 \text{ mL } NO \left(\dfrac{1 \text{ L } NO}{1000 \text{ mL } NO} \right) \left(\dfrac{1 \text{ mol } NO}{22.4 \text{ L } NO} \right) = 0.00679 \text{ mol } NO$

   $0.00679 \text{ mol } NO \left(\dfrac{3 \text{ mol } Cu}{2 \text{ mol } NO} \right) \left(\dfrac{63.5 \text{ g } Cu}{1 \text{ mol } Cu} \right) = 0.647 \text{ g } Cu$

11.67 mass of gas = 90.88 g - 90.24 g = 0.64 g

$$V_{STP} = 0.240 \text{ L} \left(\frac{273 \text{ K}}{298 \text{ K}} \right) \left(\frac{1.00 \text{ atm}}{1.00 \text{ atm}} \right) = 0.220 \text{ L} \quad \text{(at STP)}$$

$$0.220 \text{ L} \left(\frac{1 \text{ mol}}{22.4 \text{ L}} \right) = 0.00982 \text{ mol}$$

$$\text{molar mass} = \frac{0.64 \text{ g}}{0.00982 \text{ mol}} = 65 \text{ g/mol}$$

11.68 mass of gas = 4.79 g - 1.05 g = 3.74 g

$$V_{STP} = 0.600 \text{ L} \left(\frac{273 \text{ K}}{303 \text{ K}} \right) \left(\frac{762 \text{ torr}}{760 \text{ torr}} \right) = 0542 \text{ L} \quad \text{(at STP)}$$

$$0.542 \text{ L} \left(\frac{1 \text{ mol}}{22.4 \text{ L}} \right) = 0.0242 \text{ mol}$$

$$\text{molar mass} = \frac{3.74 \text{ g}}{0.0242 \text{ mol}} = 155 \text{ g/mol}$$

11.69 $P_T = 125 \text{ torr} + 225 \text{ torr} + 355 \text{ torr} = 705 \text{ torr}$

11.70 total moles = 0.250 mol + 0.370 mol + 0.155 mol = 0.775 mol

$$P_{O_2} = \frac{0.250}{0.775}(755 \text{ torr}) = 244 \text{ torr}$$

$$P_{N_2} = \frac{0.370}{0.775}(755 \text{ torr}) = 360 \text{ torr}$$

$$P_{CO_2} = \frac{0.155}{0.775}(755 \text{ torr}) = 151 \text{ torr}$$

11.71 $P_{O_2} = (0.20)760 \text{ torr} = 150 \text{ torr}$

11.72 $5.00 \text{ g CO}_2 \left(\frac{1 \text{ mol CO}_2}{44.0 \text{ g CO}_2} \right) = 0.114 \text{ mol CO}_2$

$$4.00 \text{ g CO} \left(\frac{1 \text{ mol CO}}{28.0 \text{ g CO}} \right) = 0.143 \text{ mol CO}$$

$$3.75 \text{ g Ne} \left(\frac{1 \text{ mol Ne}}{20.2 \text{ g Ne}} \right) = 0.186 \text{ mol Ne}$$

$$= 0.443 \text{ mol total}$$

11.72 (continued)

$$P_{CO_2} = \frac{0.114}{0.443}(1.35 \text{ atm}) = 0.347 \text{ atm}$$

$$P_{CO} = \frac{0.143}{0.443}(1.35 \text{ atm}) = 0.436 \text{ atm}$$

$$P_{Ne} = \frac{0.186}{0.443}(1.35 \text{ atm}) = 0.567 \text{ atm}$$

11.73 $10.00 \text{ g Ne}\left(\dfrac{1 \text{ mol Ne}}{20.2 \text{ g Ne}}\right) = 0.495 \text{ mol Ne}$

$10.0 \text{ g Ar}\left(\dfrac{1 \text{ mol Ar}}{39.9 \text{ g Ar}}\right) = 0.251 \text{ mol Ar}$

$10.0 \text{ g CO}_2\left(\dfrac{1 \text{ mol CO}_2}{44.0 \text{ g CO}_2}\right) = 0.227 \text{ mol CO}_2$

$\overline{}$

$= 0.973 \text{ mol total}$

$$P_{Ne} = \frac{0.495}{0.973}(750 \text{ torr}) = 382 \text{ torr}$$

$$P_{Ar} = \frac{0.251}{0.973}(750 \text{ torr}) = 193 \text{ torr}$$

$$P_{CO_2} = \frac{0.227}{0.973}(750 \text{ torr}) = 175 \text{ torr}$$

11.74 $P_{O_2} = 608 \text{ torr}\left(\dfrac{1.00 \text{ L}}{2.00 \text{ L}}\right) = 304 \text{ torr}$

$P_{N_2} = 732 \text{ torr}\left(\dfrac{1.00 \text{ L}}{2.00 \text{ L}}\right) = 366 \text{ torr}$

$\overline{}$

$P_{total} = 670 \text{ torr}$

11.75 $P_{O_2} = 762.3 \text{ torr} - 18.7 \text{ torr} = 743.6 \text{ torr}$

11.76 P_{H_2} = 752.7 torr - 15.5 torr = 737.2 torr

$$275 \text{ mL} \left(\frac{273 \text{ K}}{291 \text{ K}} \right) \left(\frac{737.2 \text{ torr}}{760.0 \text{ torr}} \right) = 250 \text{ mL} \ (2.50 \times 10^2 \text{ mL})$$

$$250 \text{ mL} \left(\frac{1 \text{ L}}{1000 \text{ mL}} \right) \left(\frac{1 \text{ mol}}{22.4 \text{ L}} \right) = 0.0112 \text{ mol}$$

11.77 (a) Gaseous molecules are relatively far apart and are in constant motion. Because of the space between molecules, gases are readily compressed. As a result of their constant motions, gaseous molecules will spread out if the volume to which they are confined increases.

(b) Because of the large amount of space between molecules, the ratio of mass to volume (the density) is low.

(c) Molecules are in constant motion, moving freely in all directions. As a result, they fill their container uniformly.

(d) Because each component of a mixture of gases fills its container uniformly, mixtures of molecules are also uniform.

(e) Gaseous molecules collide with all walls of their containers. Since they move in all directions, they create the same pressure on all walls.

11.78 (a) Boyle's Law: Pressure and volume are inversely proportional.

(b) Gay-Lussac's Law: Pressure and absolute temperature are directly proportional.

(c) Charles's Law: Volume and absolute temperature are directly proportional.

(d) Avogadro's Law: Volume and number of moles are directly proportional.

11.79 When a gas bubble is released at a depth of 50 meters, it is under the pressure of the column of water that is above it. However, it ascends because its density is less than that of the water surrounding it. As the bubble ascends, the surrounding pressure from the water above it decreases, which allows the bubble to expand (in accord with Boyle's Law). This expansion causes the bubble to ascend more rapidly. Thus, as a bubble ascends, it gets larger and accelerates towards toward the surface.

11.80 A balloon filled with helium ascends in the air, because it is less dense than the air surrounding it. As it ascends to higher elevations, the atmospheric pressure surrounding the balloon decreases, causing the balloon to expand. This phenomenon is an illustration of Boyle's Law.

11.81 The temperature of the water is generally cooler than that of the surroundings. Thus, when the flotation device comes in contact with the water, the gas inside cools and its pressure decreases (in accord with Gay-Lussac's Law). The best way to combat this is to add more air while the flotation device is in contact with the water. (If you try this, *do not use an electrically-driven air pump!* Addition of air using a manual pump or your mouth is safe.)

11.82 (a) If an aerosol can is thrown in a fire, the pressure of the contents increases (in accord with Gay-Lussac's Law). When the pressure becomes high enough, the can will rupture (explode).

11.82 (continued)

(b) The checked luggage compartment in an airplane may not be pressurized. Thus, an aerosol can placed in the luggage compartment may experience a decreased pressure on the outside. It is possible that the increased pressure difference between the inside and outside of the can may cause it to rupture, thereby releasing its contents into the baggage. The tendency to expand under reduced pressure is in accord with Boyle's Law.
[Note: The danger of an aerosol can exploding in an unpressurized baggage compartment is not as serious as that of throwing a can into a fire.]

11.83 $NaHCO_3 + HCl \rightarrow NaCl + H_2O + CO_2$

$$12.6 \text{ g NaHCO}_3 \left(\frac{1 \text{ mol NaHCO}_3}{84.0 \text{ g NaHCO}_3} \right) \left(\frac{1 \text{ mol CO}_2}{1 \text{ mol NaHCO}_3} \right) \left(\frac{22.4 \text{ L CO}_2}{1 \text{ mol CO}_2} \right)$$

$$= 3.36 \text{ L CO}_2 \text{ (at STP)}$$

$$3.36 \text{ L} \left(\frac{310 \text{ K}}{273 \text{ K}} \right) = 3.82 \text{ L} \quad \text{(at 37°C)}$$

11.84 (a) $d_{CO_2} = \dfrac{44.0 \text{ g}}{22.4 \text{ L}} = 1.96 \text{ g/L}$ (b) $d_{He} = \dfrac{4.0 \text{ g}}{22.4 \text{ L}} = 0.18 \text{ g/L}$

(c) $d_{N_2} = \dfrac{28.0 \text{ g}}{22.4 \text{ L}} = 1.25 \text{ g/L}$ (d) $d_{O_2} = \dfrac{32.0 \text{ g}}{22.4 \text{ L}} = 1.43 \text{ g/L}$

(e) Since air is composed of nitrogen and oxygen, its density is between 1.25 g/L and 1.43 g/L. The density of helium (0.18 g/L) is much less than this, so a helium balloon rises in air.

(f) Carbon dioxide has a density greater than that of air, and will therefore *sink* in air. Thus, it will blanket a fire, forcing the air to rise above it, away from the source of the fire.

(g) The volume of 1 mole of a gas at 20°C, 1.25 atm is:

$$V = 22.4 \text{ L} \left(\frac{293 \text{ K}}{273 \text{ K}} \right) \left(\frac{1.00 \text{ atm}}{1.25 \text{ atm}} \right) = 19.2 \text{ L}$$

$$d = \frac{44.0 \text{ g}}{19.2 \text{ L}} = 2.29 \text{ g/L}$$

11.85 (a) The molar mass may be obtained by dividing the number of grams, g, by the number of moles, n: molar mass $= \dfrac{g}{n}$. The number of moles may also be obtained from the Ideal Gas Law: $n = \dfrac{PV}{RT}$. Substitution of this latter expression for n into the earlier expression gives:

$$\text{molar mass} = \frac{g}{\dfrac{PV}{RT}} = g \cdot \frac{RT}{PV} = \frac{gRT}{PV}$$

(b) molar mass $= \dfrac{(3.74 \text{ g})\left(\dfrac{62.4 \text{ L} \cdot \text{torr}}{\text{mol} \cdot \text{K}}\right)(303 \text{ K})}{(762 \text{ torr})(0.600 \text{ L})} = 155 \text{ g/mol}$

11.86 You need only consider the *partial* pressure of oxygen.

$$V_{STP} = 7.50 \text{ L}\left(\frac{273 \text{ K}}{328 \text{ K}}\right)\left(\frac{550 \text{ torr}}{760 \text{ torr}}\right) = 4.52 \text{ L}$$

$$4.52 \text{ L}\left(\frac{1 \text{ mol}}{22.4 \text{ L}}\right) = 0.202 \text{ mol}$$

11.87 The partial pressure of CO_2 is independent of the nitrogen present. To calculate the partial pressure of CO_2 simply substitute the volume, number of moles, and temperature into the Ideal Gas Law:

$$V = (4.00 \text{ m})(3.00 \text{ m})(2.50 \text{ m}) = 30.0 \text{ m}^3\left(\frac{10^3 \text{ L}}{1 \text{ m}^3}\right) = 3.00 \times 10^4 \text{ L}$$

$$n = 1.00 \times 10^3 \text{ g } CO_2\left(\frac{1 \text{ mol } CO_2}{44.0 \text{ g } CO_2}\right) = 22.7 \text{ mol } CO_2$$

$$(P)(3.00 \times 10^4 \text{ L}) = (22.7 \text{ mol})\left(\frac{62.4 \text{ L} \cdot \text{torr}}{\text{mol} \cdot \text{K}}\right)(290 \text{ K})$$

$$P = \frac{(22.7)(62.4)(290)\text{torr}}{(3.00 \times 10^4)} = 13.7 \text{ torr}$$

CHAPTER 12

Properties of the Liquid and Solid States

12.1 (a) trigonal planer (b) angular (c) trigonal pyramidal (d) tetrahedral (e) linear

12.2 (a) nonpolar (b) polar (c) polar (d) nonpolar (e) polar (f) nonpolar (g) nonpolar (h) polar

12.3 (a) 3 (b) 1 (c) 3 (d) 2 (e) 1 (f) 2 (g) 1

12.4 (a) 2 (b) 4 (c) 1 (d) 3

12.5 CH_4 < CH_3Cl < CH_3OH

12.6 (a) NaCl (ionic) is higher boiling than CH_3CH_2Cl (dipole-dipole), because ionic interactions are stronger than dipole-dipole interactions.
(b) CH_3OCH_3 (dipole-dipole) is higher boiling than $CH_3CH_2CH_3$ (London forces), because dipole-dipole interactions are stronger than London forces.
(c) $CH_3CH_2CH_2CH_3$ (London forces) is higher boiling than $CH_3CH_2CH_3$ (London forces), because it has a higher molecular mass.
(d) ICl (dipole-dipole) is higher boiling than Br_2 (London forces), because dipole-dipole interactions are stronger than London forces.
(e) $HOCH_2CH_2OH$ (hydrogen bonding) is higher boiling than $CH_3CH_2CH_2OH$ (hydrogen bonding), because it has two groups that hydrogen bond, rather than one.

12.7, 12.8, and 12.9

	20°C (room temperature)	- 50°C	- 100°C	100°C
(a)	liquid	solid	solid	gas
(b)	solid	solid	solid	solid
(c)	gas	liquid	liquid	gas
(d)	liquid	liquid	liquid	gas
(e)	liquid	solid	solid	gas
(f)	liquid	solid	solid	gas
(g)	liquid	solid	solid	liquid
(h)	solid	solid	solid	liquid
(i)	gas	gas	solid	gas
(j)	gas	liquid	solid	gas

12.10 $55.0 \text{ g} \left(\dfrac{4.184 \text{ J}}{\text{g·°C}} \right) 53.0°C = 12,200 \text{ J} = 12.2 \text{ kJ}$

12.11 $155 \text{ g} \left(\dfrac{0.449 \text{ J}}{\text{g·°C}} \right) 78.0°C = 5430 \text{ J} = 5.43 \text{ kJ}$

12.12 $75.0 \text{ g} \left(\dfrac{0.897 \text{ J}}{\text{g} \cdot {}^\circ\text{C}} \right) 83.0^\circ\text{C} = 5580 \text{ J} = 5.58 \text{ kJ}$

12.13 $17.4 \text{ g} \left(\dfrac{0.334 \text{ kJ}}{1 \text{ g}} \right) = 5.81 \text{ kJ}$

12.14 $7.82 \text{ mol} \left(\dfrac{6.01 \text{ kJ}}{1 \text{ mol}} \right) = 47.0 \text{ kJ}$

12.15 $525 \text{ g} \left(\dfrac{2.26 \text{ kJ}}{1 \text{ g}} \right) = 1190 \text{ kJ} \quad (1.19 \times 10^3 \text{ kJ})$

12.16 $6.73 \text{ g} \left(\dfrac{2.26 \text{ kJ}}{1 \text{ g}} \right) = 15.2 \text{ kJ}$

12.17 (a) The substance with the highest vapor pressure is the one with the lowest boiling point: A (53°C).
   (b) The substance with the lowest vapor pressure is the one with the highest boiling point: B (165°C).

12.18 (a) The one with the lowest vapor pressure has the highest boiling point: E (13 torr).
   (b) At room temperature, F is a gas, because its vapor pressure (960 torr) exceeds atmospheric pressure.
   (c) We do not have enough information to determine whether any are solids.

12.19 Water will be on top, because it is less dense.

12.20 Since water is a molecular substance, a snowflake must be a molecular crystal.

12.21 When water freezes, a rigid network of hydrogen bonds forms, creating considerable space between the molecules. Because density and volume are inversely proportional, this expansion of volume produces a decrease in density. Hence the solid is less dense than the liquid.

12.22 (a) trigonal planer (b) angular (c) trigonal pyramidal (d) tetrahedral (e) angular

12.23 (a) nonpolar
   (b) polar: dipole goes through the carbon atom in the direction from the hydrogen to the plane determined by the fluorine atoms.
   (c) polar: dipole bisects the bond angle from the sulfur to the line connecting the oxygen atoms.
   (d) nonpolar
   (e) nonpolar

12.23 (continued)
(f) polar: dipole passes through the phosphorus perpendicular to the plane determined by the chlorine atoms.

12.24 (a) 2   (b) 4   (c) 1   (d) 3

12.25 (a) and (c) only.   (e) does not hydrogen-bond because bromine is not a second-row element.

12.26 (d) KI. A substance whose melting point and boiling point are extremely high is most likely to be an ionic compound, as is the case for KI.

12.27 (a) O_2. A substance whose melting point and boiling point are extremely low is most likely to be a nonpolar substance with a relatively low molecular mass, as is the case for O_2.

12.28 Whereas nitrogen is nonpolar, carbon monoxide has a net dipole. The dipole-dipole interactions in carbon monoxide cause its boiling point to be elevated above that of nitrogen.

12.29 Ammonia has the higher boiling point. Whereas methane is a nonpolar substance, ammonia is capable of hydrogen bonding. The hydrogen bonding between ammonia molecules is responsible for its higher boiling point.

12.30 (a) LiF (ionic) has a higher boiling point than CH_3F (dipole-dipole), because ionic interactions are stronger than dipole-dipole interactions.
(b) $CH_3CH_2CH_2CH_2CH_2OH$ (hydrogen bonding) has a higher boiling point than $CH_3CH_2CH_2CH_2OH$ (hydrogen bonding), because it has the larger molecular mass.
(c) $CH_3CH_2CH_2NH_2$ (hydrogen bonding), has a higher boiling than $(CH_3)_3N$ (dipole-dipole), because hydrogen bonding is a stronger interaction than dipole-dipole interactions.
(d) PH_3 (dipole-dipole) has a higher boiling point than SiH_4 (London forces), because dipole-dipole interactions are stronger than London forces.
(e) $H_2NCH_2CH_2CH_2NH_2$ (hydrogen bonding) has a higher boiling point than $CH_3CH_2CH_2CH_2NH_2$ (hydrogen bonding), because it has two groups capable of hydrogen bonding, rather than one.

12.31 When there is frost on the ground, the air temperature is at or below the freezing point of water ($0°C$). Since that temperature is below the boiling point of butane, the fuel does not vaporize very efficiently, leading to difficulty lighting the stove.

12.32, 12.33, and 12.34

	20°C	50°C	-50°C	-100°C	100°C
(a)	solid	solid	solid	solid	solid
(b)	liquid	liquid	liquid	solid	gas
(c)	gas	gas	gas	gas	gas
(d)	solid	liquid	solid	solid	liquid
(e)	gas	gas	liquid	liquid	gas
(f)	liquid	liquid	solid	solid	liquid
(g)	solid	solid	solid	solid	liquid
(h)	gas	gas	liquid	liquid	gas

12.35 Since the aluminum has a very low specific heat, the loss of a small amount of heat is accompanied by a relatively large drop in temperature.

12.36 $73.0 \text{ g} \left(\dfrac{4.184 \text{ J}}{\text{g}\cdot°\text{C}} \right) 23.0°\text{C} = 7020 \text{ J} = 7.02 \text{ kJ}$

12.37 $94.3 \text{ g} \left(\dfrac{0.385 \text{ J}}{\text{g}\cdot°\text{C}} \right) 31.0°\text{C} = 1130 \text{ J} = 1.13 \text{ kJ}$

12.38 $75.0 \text{ g} \left(\dfrac{0.129 \text{ J}}{\text{g}\cdot°\text{C}} \right) 65.0°\text{C} = 629 \text{ J}$

12.39 (a) heat of fusion (b) specific heat (c) heat of vaporization

12.40 (a) Heat of fusion.
   (b) Heat is transferred from the liquid water to the ice. This heat supplies the heat of fusion that melts the ice. As heat is lost from the liquid, its temperature drops.
   (c) Since water has a relatively high heat of fusion, it takes a lot of heat to melt a small amount of ice. Thus, a small amount of ice can cool a large amount of liquid.

12.41 When a person perspires, excess heat from the body causes the evaporation of the perspiration. Because water has a high heat of vaporization, a relatively large amount of heat is carried off during the evaporation of a relatively small amount of water. When an individual is extremely overheated, he or she may lose enough water to become dehydrated.

12.42 $3.17 \text{ mol} \left(\dfrac{6.01 \text{ kJ}}{1 \text{ mol}} \right) = 19.1 \text{ kJ}$

12.43 $7.82 \text{ g} \left(\dfrac{0.334 \text{ kJ}}{1 \text{ g}} \right) = 2.61 \text{ kJ}$

12.44 $8.50 \text{ mol} \left(\dfrac{40.7 \text{ kJ}}{1 \text{ mol}} \right) = 346 \text{ kJ}$

12.45 $76.0 \text{ g} \left(\dfrac{2.26 \text{ kJ}}{1 \text{ g}} \right) = 172 \text{ kJ}$

12.46 Molecules from the surface of the liquid are constantly vaporizing. At the same time, vapor molecules are returning to the liquid state. These opposing processes occur at equal rates, leading to a constant vapor pressure. Like all equilibrium processes, dynamic change is occurring at the microscopic (or molecular) level, while constant properties are observed at the macroscopic (or gross) level.

12.47 (a) boiling point   (b) normal boiling point   (c) evaporation   (d) boiling

12.48 When a liquid evaporates, high energy molecules at the surface of the liquid escape into the vapor phase. Since the departing molecules are those having the highest kinetic energies, the average kinetic energy of the liquid left behind decreases. Since temperature is proportional to the average kinetic energy of a sample, evaporation is accompanied by a decrease in temperature.

12.49 toluene

12.50 (a) Substance *I*   (b) gas   (c) 78°C

12.51 (a) The viscosity of a liquid decreases with increasing temperature.
(b) Attractive interactions exist between water molecules and the glass, causing the water to creep up the sides of the glass.
(c) Liquid molecules are attracted to other liquid molecules by their intermolecular forces. Molecules in the center of the liquid are acted on by forces from all sides. Molecules at the surface are only acted on from the interior side. This draws the molecules of the liquid together in a (more-or-less) spherical shape.

12.52 (a) A crystalline solid has a definite, repeating arrangement of molecules, ions, or atoms, whereas an amorphous solid lacks such a regular pattern.
(b) ionic crystal: a sodium chloride (table salt) crystal molecular crystal: a sucrose (table sugar) crystal metallic crystal: a copper wire network crystal: a diamond

12.53 *Intramolecular forces* are those that hold atoms together within each molecule, such as chemical bonds. *Intermolecular forces* are those between neighboring molecules, that hold the molecules together in large aggregates we can see at the macroscopic level.

12.54 There are no distinct molecules in ionic substances. Instead there is some type of alternating array of ions.

12.55 Carbon tetrachloride is nonpolar, and hence we would expect it to be immiscible in water. Since methyl alcohol is polar and can hydrogen bond with water molecules, we would expect it to dissolve, or be miscible, in water. In fact, both of these predictions are correct.

12.56

$$H - \underset{\displaystyle |}{\overset{\displaystyle H}{}} \overset{\displaystyle \cdot\cdot}{\underset{\displaystyle \cdot\cdot}{O}}\!\!:$$

$$CH_3CH_2OCH_2CH_3$$

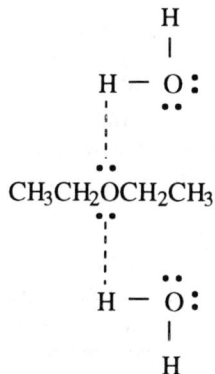

Hydrogen atoms from the water hydrogen bond to the oxygen atoms from ether molecules.

$$H - \overset{\displaystyle \cdot\cdot}{\underset{\displaystyle |}{O}}\!\!:$$
$$H$$

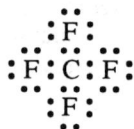

12.57 (a) $:\!\overset{\displaystyle :F:}{\underset{\displaystyle :F:}{F\!:\!C\!:\!F}}\!:$ tetrahedral, nonpolar

(b) $H\!:\!C\!:\!:\!:\!N\!:$ linear, polar

(c) $:\!\overset{\displaystyle}{\underset{\displaystyle :F:}{F\!:\!P\!:\!F}}\!:$ trigonal pyramidal, polar

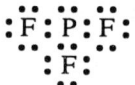

(d) $H\!:\!\overset{\displaystyle :\ddot{C}l:}{\underset{\displaystyle :\ddot{C}l:}{C}}\!:\!H$ tetrahedral, polar

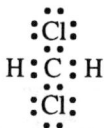

(e) $:\!\overset{\displaystyle :O:}{\underset{\displaystyle}{O}}\!:\!S\!:\!\ddot{O}\!:$ trigonal planar, nonpolar

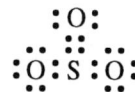

12.58 (a) boiling point: $F_2 < Cl_2 < Br_2$
(b) vapor pressure: $Br_2 < Cl_2 < F_2$
(c) molecular mass: $F_2 < Cl_2 < Br_2$
(d) London forces

12.59 Fluorine atoms are capable of engaging in hydrogen bonding with the hydrogen atoms from water. Thus, fluoroethane hydrogen bonds with water, rendering it soluble. The other halogens (Cl, Br, and I) do not engage in hydrogen bonding. Thus, chloroethane, bromoethane, and iodoethane are not soluble in water.

12.60 The surface molecules of all liquids have an attraction for molecules at the interior of the liquid, creating a surface tension. This causes liquids to form spherical droplets. However, water molecules also have a strong attraction for the atoms in glass. For water, the attraction to the glass is greater than the attraction to the interior of the liquid, and water molecules creep up the walls of a glass tube, leading to a concave meniscus. For mercury, the attraction of atoms to the glass is negligible, and mercury forms a spherical, convex meniscus.

12.61 Since the metals are colder initially than the water, heat must flow from the water to the metals. Furthermore, for each metal, the heat gained by the metal must equal the heat lost by the water. Since the water that cools the most loses more heat, it follows that the metal that comes to equilibrium at the *colder* temperature must absorb more heat. Furthermore, the metal that comes to equilibrium at the colder temperature has changed temperature less. Both the absorption of more heat and the smaller temperature change are consistent with that metal having the higher specific heat. Since metal *B* comes to equilibrium at the colder temperature, it must have the higher specific heat.

12.62 *trans*-1,2-Dichloroethene is nonpolar, because the dipoles from the C-Cl bonds cancel one another. *cis*-1,2-Dichloroethene is a polar molecule, because the angle between the dipoles leads to a net nonzero dipole (in a fashion similar to the nonzero dipole in water). Since the polar molecule would be expected to have the higher boiling point, *cis*-1,2-dichloroethene must have a boiling point of 60°C, while *trans*-1,2-dichloroethene must have the 48°C boiling point.

12.63 First calculate the volume of the room:

$$V = (5.00 \text{ m})(4.00 \text{ m})(2.50 \text{ m}) = 50.0 \text{ m}^3 \left(\frac{1000 \text{ L}}{1 \text{ m}^3} \right) = 5.00 \times 10^4 \text{ L}$$

If the room described is saturated with mercury vapor, the conditions of volume, temperature, and pressure are: $V = 5.00 \times 10^4$ L, $T = 25°C = 298$ K, and $P = 2.5 \times 10^{-6}$ atm. Converting this volume to STP will enable us to calculate the moles of mercury:

$$V_{STP} = 5.0 \times 10^4 \text{ L} \left(\frac{273 \text{ K}}{298 \text{ K}} \right) \left(\frac{2.5 \times 10^{-6} \text{ atm}}{1.00 \text{ atm}} \right) = 0.115 \text{ L}$$

$$0.115 \text{ L} \left(\frac{1 \text{ mol}}{22.4 \text{ L}} \right) = 0.00511 \text{ mol (of mercury vapor)}$$

The volume of mercury liquid that equals this quantity is:

$$0.00511 \text{ mol Hg} \left(\frac{200.6 \text{ g Hg}}{1 \text{ mol Hg}} \right) \left(\frac{1 \text{ mL Hg}}{13.6 \text{ g Hg}} \right) = 0.0754 \text{ mL Hg}$$

12.63 (continued)

Assuming approximately 25 drops/mL of liquid mercury, the number of drops equals:

$$0.0754 \text{ mL Hg} \left(\frac{25 \text{ drops Hg}}{1 \text{ mL}} \right) = 1.88 \text{ drops}$$

Approximately 2 drops of liquid mercury is enough to saturate a typical room with mercury vapor!

12.64 Warm the ice to 0°C: $36 \text{ g} \left(\dfrac{2.0 \text{ J}}{\text{g·°C}} \right) 25°C = 1800 \text{ J} = 1.8 \text{ kJ}$

Melt the ice at 0°C: $36 \text{ g} \left(\dfrac{0.334 \text{ kJ}}{\text{g}} \right) = 12 \text{ kJ}$

Warm the water to 45°C: $36 \text{ g} \left(\dfrac{4.184 \text{ J}}{\text{g·°C}} \right) 21°C = 3200 \text{ J} = 3.2 \text{ kJ}$

Total heat required: 1.8 kJ + 12 kJ + 3.2 kJ = 17 kJ

Most of the heat goes to melting the ice (heat of fusion).

12.65 To begin, calculate the heat required to melt all of the ice at 0°C:

$$20 \text{ g} \left(\frac{80 \text{ cal}}{\text{g}} \right) = 1600 \text{ cal}$$

Next calculate the drop in temperature that would result from the loss of 1600 cal from the 80 g of liquid at 70°C:

$$1600 \text{ cal} = 80 \text{ g} \left(\frac{1 \text{ cal}}{\text{g·°C}} \right) \Delta T$$

$\Delta T = 20°C$

The 80 g of water at 70°C will cool to 50°C by the melting of the ice. Now we must find out what the final temperature would be if 20 g of water at 0°C were mixed with 80 g of water at 50°C. The heat lost by the cooling water would equal the heat gained by the warming water. Let T equal the final temperature of the water.

$$\text{Heat lost} = 80 \text{ g} \left(\frac{1 \text{ cal}}{\text{g·°C}} \right) (50 - T)$$

$$\text{Heat gained} = 20 \text{ g} \left(\frac{1 \text{ cal}}{\text{g·°C}} \right) (T - 0)$$

Since the heat lost equals the heat gained, we may set these equal and solve for T: 80

$$\text{g} \left(\frac{1 \text{ cal}}{\text{g·°C}} \right) (50 - T) = 20 \text{ g} \left(\frac{1 \text{ cal}}{\text{g·°C}} \right) (T - 0)$$

$4000 - 80T = 20 T$

$4000 = 100 T$

$40°C = T$

The final temperature of the water will be 40°C.

CHAPTER 13

Solutions

13.1 Three moles. One mole of $MgCl_2$ separates into one mole of Mg^{2+} and two moles of Cl^-.

13.2 32.6 g --11.2 g = 21.4 g (dissolves in 40.0 mL H_2O)

$$100 \text{ mL } H_2O \left(\frac{21.4 \text{ g}}{40.0 \text{ mL } H_2O} \right) = 53.5 \text{ g}$$

The solubility is 53.5 g/100 mL H_2O.

13.3 AgCl , (i); $Ca(OH)_2$, (ss); $HgCl_2$, (s); NaCl , (vs); KI , (vs); $C_{12}H_{22}O_{11}$, (vs).

13.4 (a) $444 \text{ g NaOH} \left(\dfrac{1 \text{ mol NaOH}}{40.0 \text{ g NaOH}} \right) = 11.1 \text{ mol NaOH}$

$\text{molarity} = \dfrac{11.1 \text{ mol NaOH}}{5.00 \text{ L soln}} = 2.22 \text{ M NaOH}$

(b) $4.04 \text{ g KNO}_3 \left(\dfrac{1 \text{ mol KNO}_3}{101.1 \text{ g KNO}_3} \right) = 0.0400 \text{ mol KNO}_3$

$\text{molarity} = \dfrac{0.0400 \text{ mol KNO}_3}{0.200 \text{ L soln}} = 0.200 \text{ M KNO}_3$

(c) $39.0 \text{ g NaBr} \left(\dfrac{1 \text{ mol NaBr}}{102.9 \text{ g NaBr}} \right) = 0.379 \text{ mol NaBr}$

$\text{molarity} = \dfrac{0.379 \text{ mol NaBr}}{0.500 \text{ L soln}} = 0.758 \text{ M NaBr}$

(d) $2.22 \text{ g CaCl}_2 \left(\dfrac{1 \text{ mol CaCl}_2}{111.1 \text{ g CaCl}_2} \right) = 0.0200 \text{ mol CaCl}_2$

$\text{molarity} = \dfrac{0.0200 \text{ mol CaCl}_2}{0.0800 \text{ L soln}} = 0.250 \text{ M CaCl}_2$

(e) $0.385 \text{ g MgBr}_2 \left(\dfrac{1 \text{ mol MgBr}_2}{184.1 \text{ g MgBr}_2} \right) = 0.00209 \text{ mol MgBr}_2$

$\text{molarity} = \dfrac{0.00209 \text{ mol MgBr}_2}{0.00500 \text{ L soln}} = 0.418 \text{ M MgBr}_2$

13.4 (continued)

(f) $7.50 \text{ g KI} \left(\dfrac{1 \text{ mol KI}}{166.0 \text{ g KI}} \right) = 0.0452 \text{ mol KI}$

$\text{molarity} = \dfrac{0.0452 \text{ mol KI}}{0.0400 \text{ L soln}} = 1.13 \text{ M KI}$

(g) $4.36 \text{ g Na}_2\text{SO}_4 \left(\dfrac{1 \text{ mol Na}_2\text{SO}_4}{142.1 \text{ g Na}_2\text{SO}_4} \right) = 0.0307 \text{ mol Na}_2\text{SO}_4$

$\text{molarity} = \dfrac{0.0307 \text{ mol Na}_2\text{SO}_4}{0.350 \text{ L soln}} = 0.0877 \text{ M Na}_2\text{SO}_4$

(h) $12.3 \text{ g Ca(NO}_3)_2 \left(\dfrac{1 \text{ mol Ca(NO}_3)_2}{164.1 \text{ g Ca(NO}_3)_2} \right) = 0.0750 \text{ mol Ca(NO}_3)_2$

$\text{molarity} = \dfrac{0.0750 \text{ mol Ca(NO}_3)_2}{0.640 \text{ L soln}} = 0.117 \text{ M Ca(NO}_3)_2$

(i) $2.65 \text{ g CuSO}_4 \left(\dfrac{1 \text{ mol CuSO}_4}{159.6 \text{ g CuSO}_4} \right) = 0.0166 \text{ mol CuSO}_4$

$\text{molarity} = \dfrac{0.0166 \text{ mol CuSO}_4}{0.0450 \text{ L soln}} = 0.369 \text{ M CuSO}_4$

(j) $6.25 \text{ g NH}_4\text{Cl} \left(\dfrac{1 \text{ mol NH}_4\text{Cl}}{53.5 \text{ g NH}_4\text{Cl}} \right) = 0.117 \text{ mol NH}_4\text{Cl}$

$\text{molarity} = \dfrac{0.117 \text{ mol NH}_4\text{Cl}}{0.0350 \text{ L soln}} = 3.34 \text{ M NH}_4\text{Cl}$

13.5 (a) $3.00 \text{ L soln} \left(\dfrac{4.00 \text{ mol KOH}}{\text{L soln}} \right) = 12.0 \text{ mol KOH}$

(b) $1.50 \text{ L soln} \left(\dfrac{2.40 \text{ mol LiNO}_3}{\text{L soln}} \right) = 3.60 \text{ mol LiNO}_3$

(c) $0.500 \text{ L soln} \left(\dfrac{0.600 \text{ mol CaCl}_2}{\text{L soln}} \right) = 0.300 \text{ mol CaCl}_2$

(d) $0.0450 \text{ L soln} \left(\dfrac{1.60 \text{ mol FeCl}_3}{\text{L soln}} \right) = 0.0720 \text{ mol FeCl}_3$

(e) $0.150 \text{ L soln} \left(\dfrac{0.460 \text{ mol CuSO}_4}{\text{L soln}} \right) = 0.0690 \text{ mol CuSO}_4$

13.5 (continued)

(f) $0.750 \text{ L soln} \left(\dfrac{0.0440 \text{ mol MgBr}_2}{\text{L soln}} \right) = 0.0330 \text{ mol MgBr}_2$

(g) $0.00350 \text{ L soln} \left(\dfrac{3.00 \text{ mol Na}_2\text{CrO}_4}{\text{L soln}} \right) = 0.0105 \text{ mol Na}_2\text{CrO}_4$

(h) $0.0250 \text{ L soln} \left(\dfrac{0.645 \text{ mol Ba(NO}_3)_2}{\text{L soln}} \right) = 0.0161 \text{ mol Ba(NO}_3)_2$

(i) $0.0020 \text{ L soln} \left(\dfrac{12 \text{ mol HCl}}{\text{L soln}} \right) = 0.024 \text{ mol HCl}$

(j) $0.025 \text{ L soln} \left(\dfrac{18 \text{ mol H}_2\text{SO}_4}{\text{L soln}} \right) = 0.45 \text{ mol H}_2\text{SO}_4$

13.6 See the answers to problem 13.5 for the calculation of the moles of solute. To prepare each solution, dissolve the calculated mass of solute shown here *in sufficient water to achieve the total volume indicated in parentheses.*

(a) $12.0 \text{ mol KOH} \left(\dfrac{56.1 \text{ g KOH}}{1 \text{ mol KOH}} \right) = 673 \text{ g KOH (in 3.00 L)}$

(b) $3.60 \text{ mol LiNO}_3 \left(\dfrac{68.9 \text{ g LiNO}_3}{1 \text{ mol LiNO}_3} \right) = 248 \text{ g LiNO}_3 \text{ (in 1.50 L)}$

(c) $0.300 \text{ mol CaCl} \left(\dfrac{111.1 \text{ g CaCl}_2}{1 \text{ mol CaCl}_2} \right) = 33.3 \text{ g CaCl}_2 \text{ (in 0.500 L)}$

(d) $0.0720 \text{ mol FeCl}_3 \left(\dfrac{162.3 \text{ g FeCl}_3}{1 \text{ mol FeCl}_3} \right) = 11.7 \text{ g FeCl}_3 \text{ (in 45.0 mL)}$

(e) $0.0690 \text{ mol CuSO}_4 \left(\dfrac{159.6 \text{ g CuSO}_4}{1 \text{ mol CuSO}_4} \right) = 11.0 \text{ g CuSO}_4 \text{ (in 0.150 L)}$

(f) $0.0330 \text{ mol MgBr}_2 \left(\dfrac{184.1 \text{ g MgBr}_2}{1 \text{ mol MgBr}_2} \right) = 6.08 \text{ g MgBr}_2 \text{ (in 0.750 L)}$

(g) $0.0105 \text{ mol Na}_2\text{CrO}_4 \left(\dfrac{162.0 \text{ g Na}_2\text{CrO}_4}{1 \text{ mol Na}_2\text{CrO}_4} \right) = 1.70 \text{ g Na}_2\text{CrO}_4 \text{ (in 3.50 mL)}$

(h) $0.0161 \text{ mol Ba(NO}_3)_2 \left(\dfrac{261.3 \text{ g Ba(NO}_3)_2}{1 \text{ mol Ba(NO}_3)_2} \right) = 4.21 \text{ g Ba(NO}_3)_2 \text{ (in 25.0 mL)}$

13.7 (a) $0.50 \text{ mol NaOH}\left(\dfrac{1 \text{ L soln}}{6.0 \text{ mol NaOH}}\right) = 0.083 \text{ L soln (or 83 mL of soln)}$

(b) $2.5 \text{ mol NH}_3\left(\dfrac{1 \text{ L soln}}{15 \text{ mol NH}_3}\right) = 0.17 \text{ L soln (or } 1.7 \times 10^2 \text{ mL soln)}$

(c) $0.125 \text{ mol KCl}\left(\dfrac{1 \text{ L soln}}{2.45 \text{ mol KCl}}\right) = 0.0510 \text{ L soln (or 51.0 mL soln)}$

(d) $0.0575 \text{ mol H}_2\text{C}_2\text{O}_4\left(\dfrac{1 \text{ L soln}}{0.785 \text{ mol H}_2\text{C}_2\text{O}_4}\right) = 0.0732 \text{ L soln (or 73.2 mL soln)}$

(e) $0.00350 \text{ mol CaCl}_2\left(\dfrac{1 \text{ L soln}}{0.120 \text{ mol CaCl}_2}\right) = 0.0292 \text{ L soln (or 29.2 mL soln)}$

13.8 (a) $V_{\text{con}} = 500 \text{ mL soln}\left(\dfrac{3.0 \text{ M}}{12 \text{ M}}\right) = 125 \text{ mL soln}$
Dilute 125 mL of 12 M HCl to 500 mL.

(b) $V_{\text{con}} = 250 \text{ mL soln}\left(\dfrac{1.0 \text{ M}}{18 \text{ M}}\right) = 14 \text{ mL soln}$
Dilute 14 mL of 18 M H_2SO_4 to 250 mL.

(c) $V_{\text{con}} = 75 \text{ mL soln}\left(\dfrac{0.60 \text{ M}}{3.0 \text{ M}}\right) = 15 \text{ mL soln}$
Dilute 15 mL of 3.0 M NaOH to 75 mL.

(d) $V_{\text{con}} = 150 \text{ mL soln}\left(\dfrac{0.10 \text{ M}}{3.0 \text{ M}}\right) = 5.0 \text{ mL soln}$
Dilute 5 mL of 3.0 M KI to 150 mL.

(e) $V_{\text{con}} = 5.0 \text{ L soln}\left(\dfrac{0.25 \text{ M}}{15 \text{ M}}\right) = 0.083 \text{ L soln} = 83 \text{ mL soln}$
Dilute 83 mL of 15 M NH_3 to 5.0 L.

13.9 (a) $M_{\text{dil}} = \dfrac{(75 \text{ mL soln})(18 \text{ M})}{(350 \text{ mL soln})} = 3.9 \text{ M}$

(b) $M_{\text{dil}} = \dfrac{(25 \text{ mL soln})(2.4 \text{ M})}{(750 \text{ mL soln})} = 0.080 \text{ M}$

(c) $M_{\text{dil}} = \dfrac{(10.0 \text{ mL soln})(1.25 \text{ M})}{(60.0 \text{ mL soln})} = 0.208 \text{ M}$

13.9 (continued)

(d) $M_{dil} = \dfrac{(0.0350 \text{ L soln})(2.00 \text{ M})}{(0.500 \text{ L soln})} = 0.140 \text{ M}$

(e) $M_{dil} = \dfrac{(5.00 \text{ mL soln})(6.00 \text{ M})}{(1250 \text{ mL soln})} = 0.0240 \text{ M}$

13.10 (a) $\dfrac{5.0 \text{ mL}}{30.0 \text{ mL}} \times 100\% = 17\%$

(b) $\dfrac{15 \text{ mL}}{60 \text{ mL}} \times 100\% = 25\%$

(c) $\dfrac{3.0 \text{ g}}{60.0 \text{ g}} \times 100\% = 5.0\%$

(d) $\dfrac{14 \text{ g}}{250 \text{ g}} \times 100\% = 5.6\%$

13.11 (a) 7.0% of 500 mL: $(0.070)(500 \text{ mL}) = 35 \text{ mL}$
Add 35 mL of acetic acid to enough water to make 500 mL of solution.
(b) 7.0% of 500 g: $(0.070)(500 \text{ g}) = 35 \text{ g}$
Add 35 g of acetic acid to 465 g of water.
(c) 12% of 250 mL: $(0.12)(250 \text{ mL}) = 30 \text{ mL} (3.0 \times 10^1 \text{ mL})$
Add 30 mL of ethyl alcohol to enough water to make 250 mL of solution.
(d) 6.0% of 750 g: $(0.060)(750 \text{ g}) = 45 \text{ g}$
Add 45 g of H_3BO_3 to 705 g of water.
(e) 2.0% of 25 g: $(0.020)(25 \text{ g}) = 0.50 \text{ g}$
Add 0.50 g of $NaHCO_3$ to 24.5 g of water.

13.12 $\dfrac{2.1 \times 10^{-3} \text{ g}}{15 \times 10^3 \text{ g}} \times 10^6 \text{ ppm} = 0.14 \text{ ppm}$

No. This level exceeds the recommended limit of 0.05 ppm.

13.13 $0.0645 \text{ L soln} \left(\dfrac{0.125 \text{ mol } H_2SO_4}{\text{L soln}} \right)\left(\dfrac{1 \text{ mol } BaSO_4}{1 \text{ mol } H_2SO_4} \right)\left(\dfrac{233.4 \text{ g } BaSO_4}{1 \text{ mol } BaSO_4} \right) = 1.88 \text{ g } BaSO_4$

13.14 $0.0550 \text{ L soln} \left(\dfrac{0.250 \text{ mol } AgNO_3}{\text{L soln}} \right)\left(\dfrac{1 \text{ mol } Ag_2S}{2 \text{ mol } AgNO_3} \right)\left(\dfrac{247.9 \text{ g } Ag_2S}{1 \text{ mol } Ag_2S} \right)$

$= 1.70 \text{ g } Ag_2S$

13.15 $3.55 \text{ g CaCO}_3 \left(\dfrac{1 \text{ mol CaCO}_3}{100.1 \text{ g CaCO}_3} \right) \left(\dfrac{2 \text{ mol HNO}_3}{1 \text{ mol CaCO}_3} \right) = 0.0709 \text{ mol HNO}_3$

$0.0709 \text{ mol HNO}_3 \left(\dfrac{1 \text{ L soln}}{0.275 \text{ mol HNO}_3} \right) = 0.258 \text{ L soln} \ \ (258 \text{ mL soln})$

13.16 $0.0400 \text{ L soln} \left(\dfrac{0.350 \text{ mol Na}_3\text{PO}_4}{\text{L soln}} \right) \left(\dfrac{3 \text{ mol Pb(NO}_3)_2}{2 \text{ mol Na}_3\text{PO}_4} \right) = 0.0210 \text{ mol Pb(NO}_3)_2$

$0.0210 \text{ mol Pb(NO}_3)_2 \left(\dfrac{1 \text{ L soln}}{0.125 \text{ mol Pb(NO}_3)_2} \right) = 0.168 \text{ L soln} \ \ (168 \text{ mL soln})$

13.17 (a) $0.0160 \text{ L soln} \left(\dfrac{0.150 \text{ mol HCl}}{\text{L soln}} \right) \left(\dfrac{1 \text{ mol H}_2}{2 \text{ mol HCl}} \right) \left(\dfrac{22.4 \text{ L H}_2}{1 \text{ mol H}_2} \right) = 0.0269 \text{ L H}_2 \text{ (at STP)}$

(b) $0.0269 \text{ L H}_2 \left(\dfrac{308 \text{ K}}{273 \text{ K}} \right) \left(\dfrac{1.00 \text{ atm}}{1.15 \text{ atm}} \right) = 0.0264 \text{ L H}_2 \ \ (26.4 \text{ mL H}_2)$

13.18 $0.0350 \text{ L soln} \left(\dfrac{0.440 \text{ mol HCl}}{\text{L soln}} \right) \left(\dfrac{1 \text{ mol NaOH}}{1 \text{ mol HCl}} \right) = 0.0154 \text{ mol NaOH}$

$\text{molarity} = \dfrac{0.0154 \text{ mol NaOH}}{0.0250 \text{ L soln}} = 0.616 \text{ M NaOH}$

13.19 $0.0250 \text{ L soln} \left(\dfrac{0.210 \text{ mol KOH}}{\text{L soln}} \right) \left(\dfrac{1 \text{ mol H}_3\text{PO}_4}{3 \text{ mol KOH}} \right) = 0.00175 \text{ mol H}_3\text{PO}_4$

$\text{molarity} = \dfrac{0.00175 \text{ mol H}_3\text{PO}_4}{0.0140 \text{ L soln}} = 0.125 \text{ M H}_3\text{PO}_4$

13.20 (a) $\Delta T = \left(\dfrac{1.86°\text{C}}{\text{m}} \right) 1.20 \text{ m} = 2.23°\text{C}$

$T_f = 0.00°\text{C} - 2.23°\text{C} = -2.23°\text{C}$

(b) $\Delta T = \left(\dfrac{3.9°\text{C}}{\text{m}} \right) 2.34 \text{ m} = 9.1°\text{C}$

$T_f = 16.6°\text{C} - 9.1°\text{C} = 7.5°\text{C}$

(c) $\Delta T = \left(\dfrac{40.0°\text{C}}{\text{m}} \right) 0.670 \text{ m} = 26.8°\text{C}$

$T_f = 17.6°\text{C} - 26.8°\text{C} = -9.2°\text{C}$

13.20 (continued)

(d) $\Delta T = \left(\dfrac{4.9°C}{m} \right) 0.23\ m = 1.1°C$

$T_f = 5.5°C - 1.1°C = 4.4°C$

(e) $\Delta T = \left(\dfrac{6.8°C}{m} \right) 0.78\ m = 5.3°C$

$T_f = 80.2°C - 5.3°C = 74.9°C$

13.21 (a)   The water will flow to the right.

(b)   Because 0.9% NaCl is isotonic with 5% glucose and 5% NaCl is more concentrated than 0.9% NaCl, water will flow to the left.

13.22 It does not settle out upon standing, and it is milky in appearance.

13.23 (a)   A solution is any homogeneous mixture.

(b)   2, 3, and 4 are solutions.

(c)   A *liquid solution* is one prepared by dissolving a solid, a liquid, or a gas in a liquid.

(d)   The *solute* is the substance being dissolved, and the *solvent* is the liquid in which it is dissolved.

(e)   An *aqueous solution* is a solution for which the solvent is water.

(f)   Solutions are readily dispensed, and their solvents provide a medium in which the reactant ions and molecules are able to collide with one another.

13.24 (a)   The *solubility* of a substance is its maximum concentration in a given solvent.

(b)   Yes.

(c)   The solubilities of *most* solids increase with increasing temperature. The solubilities of *all* gases decrease with increasing temperature.

(d)   Polarity. Like dissolves like. Polar solvents best dissolve polar solutes, whereas nonpolar solvents best dissolve nonpolar solutes.

(e)   Yes. It hydrogen bonds to the water molecules.

(f)   No. A nonpolar solute is generally insoluble in a polar solvent.

(g)   The rate of dissolving increases with decreasing particle size. Stirring increases the rate of dissolving. As the temperature increases, so does the rate of dissolving.

13.25 The solubilities of gases in liquids decrease with increasing temperature. Boiling expels the dissolved gases.

13.26 9.85 g – 2.73 g = 7.12 g   (dissolves in 35.0 mL H_2O)

$100\ mL\ H_2O \left(\dfrac{7.12\ g}{35.0\ mL\ H_2O} \right) = 20.3\ g$

The solubility is 20.3 g/100 mL H_2O

13.27 *Molarity* is the number of moles per liter. To prepare one liter of 0.500 M KI, place 83.0 g KI in a 1-liter volumetric flask, add enough water to bring the volume to the 1-liter mark, and mix. (It is desirable to dissolve the solute before bringing the final volume to the 1-liter mark.)

13.28 (a) $135 \text{ g KCl} \left(\dfrac{1 \text{ mol KCl}}{74.6 \text{ g KCl}} \right) = 1.81 \text{ mol KCl}$

$\text{molarity} = \left(\dfrac{1.81 \text{ mol KCl}}{2.50 \text{ L soln}} \right) = 0.724 \text{ M KCl}$

(b) $403 \text{ g MgSO}_4 \left(\dfrac{1 \text{ mol MgSO}_4}{120.4 \text{ g MgSO}_4} \right) = 3.35 \text{ mol MgSO}_4$

$\text{molarity} = \left(\dfrac{3.35 \text{ mol MgSO}_4}{5.25 \text{ L soln}} \right) = 0.638 \text{ M MgSO}_4$

(c) $5.00 \text{ g KBr} \left(\dfrac{1 \text{ mol KBr}}{119.0 \text{ g KBr}} \right) = 0.0420 \text{ mol KBr}$

$\text{molarity} = \left(\dfrac{0.0420 \text{ mol KBr}}{0.500 \text{ L soln}} \right) = 0.0840 \text{ M KBr}$

(d) $0.554 \text{ g NH}_3 \left(\dfrac{1 \text{ mol NH}_3}{17.0 \text{ g NH}_3} \right) = 0.0326 \text{ mol NH}_3$

$\text{molarity} = \left(\dfrac{0.0326 \text{ mol NH}_3}{0.0150 \text{ L soln}} \right) = 2.17 \text{ M NH}_3$

(e) $6.12 \text{ g C}_6\text{H}_{12}\text{O}_6 \left(\dfrac{1 \text{ mol C}_6\text{H}_{12}\text{O}_6}{180.0 \text{ g C}_6\text{H}_{12}\text{O}_6} \right) = 0.0340 \text{ mol C}_6\text{H}_{12}\text{O}_6$

$\text{molarity} = \left(\dfrac{0.0340 \text{ mol C}_6\text{H}_{12}\text{O}_6}{0.0500 \text{ L soln}} \right) = 0.680 \text{ M C}_6\text{H}_{12}\text{O}_6$

(f) $1.24 \text{ g AgNO}_3 \left(\dfrac{1 \text{ mol AgNO}_3}{169.9 \text{ g AgNO}_3} \right) = 0.00730 \text{ mol AgNO}_3$

$\text{molarity} = \left(\dfrac{0.00730 \text{ mol AgNO}_3}{0.125 \text{ L soln}} \right) = 0.0584 \text{ M AgNO}_3$

13.29 (a) $2.50 \text{ L soln} \left(\dfrac{3.50 \text{ mol NaCl}}{\text{L soln}} \right) = 8.75 \text{ mol NaCl}$

13.29 (continued)

(b) $0.375 \text{ L soln}\left(\dfrac{6.00 \text{ mol HCl}}{\text{L soln}}\right) = 2.25 \text{ mol HCl}$

(c) $0.0550 \text{ L soln}\left(\dfrac{3.00 \text{ mol NaOH}}{\text{L soln}}\right) = 0.165 \text{ mol NaOH}$

(d) $0.0236 \text{ L soln}\left(\dfrac{0.125 \text{ mol H}_2\text{SO}_4}{\text{L soln}}\right) = 0.00295 \text{ mol H}_2\text{SO}_4$

13.30   (a) $8.75 \text{ mol NaCl}\left(\dfrac{58.5 \text{ g NaCl}}{1 \text{ mol NaCl}}\right) = 512 \text{ g NaCl}$

(b) $2.25 \text{ mol HCl}\left(\dfrac{36.5 \text{ g HCl}}{1 \text{ mol HCl}}\right) = 82.1 \text{ g HCl}$

(c) $0.165 \text{ mol NaOH}\left(\dfrac{40.0 \text{ g NaOH}}{1 \text{ mol NaOH}}\right) = 6.60 \text{ g NaOH}$

(d) $0.00295 \text{ mol H}_2\text{SO}_4\left(\dfrac{98.1 \text{ g H}_2\text{SO}_4}{1 \text{ mol H}_2\text{SO}_4}\right) = 0.289 \text{ g H}_2\text{SO}_4$

13.31   (a) $2.50 \text{ L soln}\left(\dfrac{0.250 \text{ mol (NH}_4)_2\text{CO}_3}{\text{L soln}}\right)\left(\dfrac{96.0 \text{ g (NH}_4)_2\text{CO}_3}{1 \text{ mol (NH}_4)_2\text{CO}_3}\right) = 60.0 \text{ g}$
(NH$_4$)$_2$CO$_3$

(b) $0.125 \text{ L soln}\left(\dfrac{2.50 \text{ mol NH}_4\text{Cl}}{\text{L soln}}\right)\left(\dfrac{53.5 \text{ g NH}_4\text{Cl}}{1 \text{ mol NH}_4\text{Cl}}\right) = 16.7 \text{ g NH}_4\text{Cl}$

(c) $0.0375 \text{ L soln}\left(\dfrac{0.750 \text{ mol KI}}{\text{L soln}}\right)\left(\dfrac{166.0 \text{ g KI}}{1 \text{ mol KI}}\right) = 4.67 \text{ g KI}$

(d) $0.0750 \text{ L soln}\left(\dfrac{0.160 \text{ mol C}_{12}\text{H}_{22}\text{O}_{11}}{\text{L soln}}\right)\left(\dfrac{342.0 \text{ g C}_{12}\text{H}_{22}\text{O}_{11}}{1 \text{ mol C}_{12}\text{H}_{22}\text{O}_{11}}\right) = 4.10 \text{ g C}_{12}\text{H}_{22}\text{O}_{11}$

13.32   (a) $0.500 \text{ mol KCl}\left(\dfrac{1 \text{ L soln}}{2.00 \text{ mol KCl}}\right) = 0.250 \text{ L soln} = 250 \text{ mL soln}$

(b) $0.655 \text{ mol NaI}\left(\dfrac{1 \text{ L soln}}{2.50 \text{ mol NaI}}\right) = 0.262 \text{ L soln} = 262 \text{ mL soln}$

(c) $0.0435 \text{ mol H}_2\text{SO}_4\left(\dfrac{1 \text{ L soln}}{0.125 \text{ mol H}_2\text{SO}_4}\right) = 0.348 \text{ L soln} = 348 \text{ mL soln}$

13.32 (continued)

(d) $0.50 \text{ mol HCl} \left(\dfrac{1 \text{ L soln}}{12 \text{ mol HCl}} \right) = 0.042 \text{ L soln} = 42 \text{ mL soln}$

(e) $0.15 \text{ mol NaOH} \left(\dfrac{1 \text{ L soln}}{6.0 \text{ mol NaOH}} \right) = 0.025 \text{ L soln} = 25 \text{ mL soln}$

(f) $2.0 \text{ g NaOH} \left(\dfrac{1 \text{ mol NaOH}}{40.0 \text{ g NaOH}} \right) \left(\dfrac{1 \text{ L soln}}{6.0 \text{ mol NaOH}} \right) = 0.0083 \text{ L soln} = 8.3 \text{ mL soln}$

13.33 (a) $V_{con} = \dfrac{(500 \text{ mL soln})(1.0 \text{ M})}{(12 \text{ M})} = 42 \text{ mL soln}$

Dilute 42 mL of 12 M HCl to 500 mL.

(b) $V_{con} = \dfrac{(250 \text{ mL soln})(2.0 \text{ M})}{(15 \text{ M})} = 33 \text{ mL soln}$

Dilute 33 mL of 15 M HNO_3 to 250 mL.

(c) $V_{con} = \dfrac{(150 \text{ mL soln})(0.60 \text{ M})}{(18 \text{ M})} = 5.0 \text{ mL soln}$

Dilute 5.0 mL of 18 M H_2SO_4 to 150 mL.

(d) $V_{con} = \dfrac{(750 \text{ mL soln})(0.25 \text{ M})}{(6.0 \text{ M})} = 31 \text{ mL soln}$

Dilute 31 mL of 6.0 M NaOH to 750 mL.

13.34 (a) $M_{dil} = \dfrac{(75 \text{ mL soln})(15 \text{ M})}{(250 \text{ mL soln})} = 4.5 \text{ M}$

(b) $M_{dil} = \dfrac{(25 \text{ mL soln})(18 \text{ M})}{(125 \text{ mL soln})} = 3.6 \text{ M}$

(c) $M_{dil} = \dfrac{(0.0500 \text{ L soln})(2.00 \text{ M})}{(0.250 \text{ L soln})} = 0.400 \text{ M}$

(d) $M_{dil} = \dfrac{(0.015 \text{ L soln})(12 \text{ M})}{(0.50 \text{ L soln})} = 0.36 \text{ M}$

13.35 (a) $(0.040)(125 \text{ mL}) = 5.0 \text{ mL}$

Dilute 5.0 mL of isopropyl alcohol to 125 mL.

(b) $(0.050)(500 \text{ mL}) = 25 \text{ mL}$

Dilute 25 mL of acetic acid to 500 mL.

(c) $(0.050)(300 \text{ g}) = 15 \text{ g}$

Add 15 g of glucose to 285 g of water.

(d) $(0.0090)(750 \text{ g}) = 6.8 \text{ g}$

Add 6.8 g of sodium chloride to 743.2 g of water.

13.36 parts per million (ppm) = milligrams per liter (mg/L)

$$\frac{0.055 \text{ mg}}{0.750 \text{ L}} = 0.073 \text{ ppm} \qquad \text{No. This is greater than 0.002 ppm.}$$

13.37 1 ppm = 0.0001% (or 1% = 10,000 ppm). Parts per billion is the number of grams per 10^9 grams of solution. This equals the number of milligrams per 1000 L of solution (or the number of micrograms per liter of solution).

13.38 $0.0750 \text{ L soln} \left(\frac{0.240 \text{ mol MgCl}_2}{\text{L soln}} \right) \left(\frac{1 \text{ mol Mg(OH)}_2}{1 \text{ mol MgCl}_2} \right) \left(\frac{58.3 \text{ g Mg(OH)}_2}{1 \text{ mol Mg(OH)}_2} \right)$

$= 1.05 \text{ g Mg(OH)}_2$

13.39 $0.0550 \text{ L soln} \left(\frac{0.125 \text{ mol Ba(OH)}_2}{\text{L soln}} \right) \left(\frac{1 \text{ mol BaCO}_3}{1 \text{ mol Ba(OH)}_2} \right) \left(\frac{197.3 \text{ g BaCO}_3}{1 \text{ mol BaCO}_3} \right)$

$= 1.36 \text{ g BaCO}_3$

13.40 (a) $0.0750 \text{ L soln} \left(\frac{0.125 \text{ mol Na}_2\text{CO}_3}{\text{L soln}} \right) \left(\frac{1 \text{ mol CO}_2}{1 \text{ mol Na}_2\text{CO}_3} \right) \left(\frac{22.4 \text{ L CO}_2}{1 \text{ mol CO}_2} \right)$

$= 0.210 \text{ L CO}_2$ (at STP)

(b) $0.210 \text{ L CO}_2 \left(\frac{309 \text{ K}}{273 \text{ K}} \right) \left(\frac{760 \text{ torr}}{715 \text{ torr}} \right) = 0.253 \text{ L CO}_2$ (at 36°C, 715 torr)

13.41 $2.5 \text{ g Cu} \left(\frac{1 \text{ mol Cu}}{63.5 \text{ g Cu}} \right) \left(\frac{8 \text{ mol HNO}_3}{3 \text{ mol Cu}} \right) \left(\frac{1 \text{ L soln}}{15 \text{ mol HNO}_3} \right) \left(\frac{1000 \text{ mL soln}}{1 \text{ L soln}} \right) = 7.0 \text{ mL soln})$

13.42 $0.0120 \text{ L soln} \left(\frac{0.375 \text{ mol AgNO}_3}{\text{L soln}} \right) \left(\frac{1 \text{ mol K}_2\text{CrO}_4}{2 \text{ mol AgNO}_3} \right) = 0.00225 \text{ mol K}_2\text{CrO}_4$

$0.00225 \text{ mol K}_2\text{CrO}_4 \left(\frac{1 \text{ L soln}}{0.180 \text{ mol K}_2\text{CrO}_4} \right) = 0.0125 \text{ L soln}$ (12.5 mL soln)

13.43 $0.0152 \text{ L soln} \left(\frac{0.240 \text{ mol H}_2\text{SO}_4}{\text{L soln}} \right) \left(\frac{2 \text{ mol KOH}}{1 \text{ mol H}_2\text{SO}_4} \right) = 0.00730 \text{ mol KOH}$

$\text{molarity} = \frac{0.00730 \text{ mol KOH}}{0.0273 \text{ L soln}} = 0.267 \text{ M KOH}$

13.44 $0.0328 \text{ L soln} \left(\dfrac{0.225 \text{ mol KI}}{\text{L soln}} \right) \left(\dfrac{1 \text{ mol Pb(NO}_3)_2}{2 \text{ mol KI}} \right) = 0.00369 \text{ mol Pb(NO}_3)_2$

   $\text{molarity} = \dfrac{0.00369 \text{ mol Pb(NO}_3)_2}{0.0126 \text{ L soln}} = 0.293 \text{ M Pb(NO}_3)_2$

13.45 (a) $\Delta T = \left(\dfrac{1.86°C}{m} \right) 2.50 \text{ m} = 4.65°C$

   $T_f = 0.00°C - 4.65°C = -4.65°C$

   (b) $\Delta T = \left(\dfrac{2.7°C}{m} \right) 2.50 \text{ m} = 6.8°C$

   $T_f = 8.4°C - 6.8°C = 1.6°C$

   (c) $\Delta T = \left(\dfrac{11.8°C}{m} \right) 2.50 \text{ m} = 29.5°C$

   $T_f = 10.0°C - 29.5°C = -19.5°C$

13.46 (a) Fluids will enter the cell, and the cell will burst.
   (b) The cell will be unaffected.
   (c) Fluids will leave the cell, and the cell will shrivel.

13.47 (a) 0.24 m glucose   (b) water   (c) 0.24 m glucose

13.48 (a) 3   (b) 5   (c) 1   (d) 6   (e) 4   (f) 2

13.49 Salt lowers the freezing point of water, thereby melting the ice on the roads.

13.50 (a) As solvent crystallizes out of the freezing solution, the concentration increases, leading to a greater freezing point depression. Thus, the freezing point is steadily lowered as freezing continues.
   (b) As a result of the behavior described in (a), it becomes increasingly more difficult to freeze the water out of a salt water mixture as freezing progresses.

13.51 (a) Since the density of alcohol is less than 1 g/mL, it is more advantageous to report the percent volume. For example, suppose you have 100 mL of a 10% ethyl alcohol solution by volume. This solution contains 10 mL of alcohol (10% of 100 mL). Since the density of the alcohol is 0.79 g/mL, the mass of the 10 mL of alcohol is 7.9 g, and the mass of the entire solution is approximately 98 g (roughly 90 g of water and 7.9 g of alcohol). The percent of this solution by mass is approximately 8%.
   (b) If you have a solute with a density greater than the density of water, the situation is just the opposite of that for the ethyl alcohol solution. The percent by mass will give the larger number, and it would be more advantageous to label the contents with the percent by mass.

13.52 (a) 25 mg = 0.025 g

$$0.025 \text{ g } C_{15}H_{15}NO_3 \left(\frac{1 \text{ mol } C_{15}H_{15}NO_3}{257.0 \text{ g } C_{15}H_{15}NO_3} \right) = 9.7 \times 10^{-5} \text{ mol } C_{15}H_{15}NO_3$$

$$\text{molarity} = \left(\frac{9.7 \times 10^{-5} \text{ mol } C_{15}H_{15}NO_3}{0.0010 \text{ L}} \right) = 9.7 \times 10^{-2} \text{ M } C_{15}H_{15}NO_3$$

(b) The dilution formula is general and can be applied to any concentration unit, as long as the same units are used on both sides of the equation. Rather than use molarities, we will use the concentrations, C, given in mg/mL:

$$V_{con} \cdot C_{con} = V_{dil} \cdot C_{dil}$$
$$(20 \text{ mL})(25 \text{ mg/mL}) = V_{dil}(10 \text{ mg/mL})$$

$$\frac{(20 \text{ mL})(25 \text{ mg / mL})}{(10 \text{ mg / mL})} = V_{dil} = 50 \text{ mL}$$

(c) $5.0 \text{ mL} \left(\dfrac{10 \text{ mg}}{\text{mL}} \right) = 50 \text{ mg}$

13.53 [To avoid excessive error, we will use 3 significant figures and round off at the end.]
The mass of an 8-oz glass of milk is:

$$236 \text{ mL} \left(\frac{1.03 \text{ g}}{\text{mL}} \right) = 243 \text{ g}$$

If 2% is fat, the mass of fat in an 8-oz serving is:
(0.02)(243 g) = 4.86 g
Since each gram of fat is 9 calories, the number of calories from fat is:

$$4.86 \text{ g} \left(\frac{9 \text{ cal}}{\text{g}} \right) = 43.7 \text{ cal}$$

The percentage of calories from fat is:

$$\% \text{ fat} = \left(\frac{43.7 \text{ cal}}{140 \text{ cal}} \right) \times 100\% = 31.2\% \text{ (or roughly 30\%)}$$

Lowfat milk exceeds the 20% target maximum!

13.54 $\text{mass of water} = (25 \text{ m})(15 \text{ m})(2.0 \text{ m}) \left(\dfrac{1000 \text{ L}}{1 \text{ m}^3} \right) \left(\dfrac{1000 \text{ mL}}{1 \text{ L}} \right) \left(\dfrac{1.00 \text{ g}}{1 \text{ mL}} \right)$

$= 7.5 \times 10^8 \text{ g}$ (or 750 million grams of water)

$$\text{parts per million} = \left(\frac{8.0 \text{ g salt}}{7.5 \times 10^8 \text{ g water}} \right) \times 10^6 \text{ ppm} = 0.011 \text{ ppm}$$

13.55 5.0 ppm Pb = 5.0 mg Pb/L soln = 0.0050 g Pb/L soln. Simply convert the value in g/L soln to mol/L soln:

$$\text{molarity} = \frac{0.0050 \text{ g Pb}}{\text{L soln}} \left(\frac{1 \text{ mol Pb}}{207.2 \text{ g Pb}} \right) = 2.4 \times 10^{-5} \text{ M Pb}$$

13.56 (a) $0.0350 \text{ L soln} \left(\dfrac{0.300 \text{ mol Pb(NO}_3)_2}{\text{L soln}} \right) = 0.0105 \text{ mol Pb(NO}_3)_2$

$0.0650 \text{ L soln} \left(\dfrac{0.200 \text{ mol KI}}{\text{L soln}} \right) = 0.0130 \text{ mol KI}$

Assume that $Pb(NO_3)_2$ is the limiting reagent:

$0.0105 \text{ mol Pb(NO}_3)_2 \left(\dfrac{2 \text{ mol KI}}{1 \text{ mol Pb(NO}_2)_3} \right) = 0.0210 \text{ mol KI}$

There is not enough KI to react with all of the $Pb(NO_3)_2$. Thus, KI is the limiting reagent.

(b) $0.0130 \text{ mol KI} \left(\dfrac{1 \text{ mol PbI}_2}{2 \text{ mol KI}} \right) \left(\dfrac{461.0 \text{ g PbI}_2}{1 \text{ mol PbI}_2} \right) = 3.00 \text{ g PbI}_2$

13.57 (a) Since 1.00 mol NaCl contains 2.00 moles of ions, this behaves like a 2.00 m solution:

$$\Delta T = \left(\frac{1.86°C}{m} \right) 2.00 \text{ m} = 3.72°C$$

$T_f = 0.00°C - 3.72°C = -3.72°C$

(b) Since 1.00 mol $MgCl_2$ contains 3.00 moles of ions, this behaves like a 3.00 m solution:

$$\Delta T = \left(\frac{1.86°C}{m} \right) 3.00 \text{ m} = 5.58°C$$

$T_f = 0.00°C - 5.58°C = -5.58°C$

(c) If 50% of 1.00 mol of AZ dissociates, this will produce 0.50 mol A^+, 0.50 mol Z^-, and 0.50 mol of undissociated AZ. The solution produced will behave like a 1.50 m solution:

$$\Delta T = \left(\frac{1.86°C}{m} \right) 1.50 \text{ m} = 2.79°C$$

$T_f = 0.00°C - 2.79°C = -2.79°C$

13.58 One part per trillion (1 ppt) is one *nano*gram per liter: 1×10^{-9} g/L.
The number of cadmium ions in 3 mL of a 3 ppt solution is:

$$3 \text{ mL soln} \left(\frac{1 \text{ L soln}}{1000 \text{ mL soln}} \right) \left(\frac{3 \times 10^{-9} \text{ g}}{\text{L soln}} \right) \left(\frac{1 \text{ mol Cd}}{112.4 \text{ g Cd}} \right) \left(\frac{6.02 \times 10^{23} \text{ atoms Cd}}{1 \text{ mol Cd}} \right)$$

$= 5 \times 10^{10}$ atoms (approximately 50 billion atoms)

CHAPTER 14

The Reactions of Aqueous Solutions

14.1 (a) combustion   (b) double-replacement   (c) decomposition
      (d) synthesis   (e) single-replacement

14.2 (a) $Pb(NO_3)_2(aq) + BaCl_2(aq) \longrightarrow PbCl_2(s) + Ba(NO_3)_2(aq)$

      (b) $H_2SO_4(aq) + 2\ KOH(aq) \longrightarrow K_2SO_4(aq) + 2\ H_2O(\ell)$

14.3 (a) nonelectrolyte   (b) electrolyte   (c) weak electrolyte
      (d) electrolyte   (e) weak electrolyte   (f) nonelectrolyte

14.4 $HBr(g) + H_2O(\ell) \longrightarrow H_3O^+(aq) + Br^-(aq)$

14.5 1 : 1 ratio

14.6 (a) $HCl + KOH \longrightarrow KCl + H_2O$

      (b) $HNO_3 + NaOH \longrightarrow NaNO_3 + H_2O$

      (c) $2\ HBr + Ca(OH)_2 \longrightarrow CaBr_2 + 2\ H_2O$

      (d) $H_3PO_4 + 3\ KOH \longrightarrow K_3PO_4 + 3\ H_2O$

      (e) $H_2SO_4 + Ca(OH)_2 \longrightarrow CaSO_4 + 2\ H_2O$

14.7 (a) $HI + NaOH \longrightarrow NaI + H_2O$

      (b) $HBr + LiOH \longrightarrow LiBr + H_2O$

      (c) $2\ HCl + Ca(OH)_2 \longrightarrow CaCl_2 + 2\ H_2O$

      (d) $H_3PO_4 + 3\ NaOH \longrightarrow Na_3PO_4 + 3\ H_2O$

      (e) $2\ HI + Ba(OH)_2 \longrightarrow BaI_2 + 2\ H_2O$

14.8 (a) $HgO + 2\ HCl \longrightarrow HgCl_2 + H_2O$

      (b) $Na_2O + 2\ HI \longrightarrow 2\ NaI + H_2O$

      (c) $2\ HBr + CaCO_3 \longrightarrow CaBr_2 + H_2O + CO_2$

      (d) $2\ HNO_3 + MgCO_3 \longrightarrow Mg(NO_3)_2 + H_2O + CO_2$

      (e) $HNO_3 + NaHCO_3 \longrightarrow NaNO_3 + H_2O + CO_2$

      (f) $HI + LiHCO_3 \longrightarrow LiI + H_2O + CO_2$

14.9   (a) $Zn^{2+}(aq)$ and $Cl^-(aq)$ in a 1:2 ratio

      (b) $Li^+(aq)$ and $SO_4^{2-}(aq)$ in a 2:1 ratio

      (c) $Fe^{3+}(aq)$ and $NO_3^-(aq)$ in a 1:3 ratio

      (d) $NH_4^+(aq)$ and $CO_3^{2-}(aq)$ in a 2:1 ratio

      (e) $Ba^{2+}(aq)$ and $I^-(aq)$ in a 1:2 ratio

14.10   (a) $HClO_4(aq) + KOH(aq) \longrightarrow KClO_4(aq) + H_2O(\ell)$

           $H^+(aq) + ClO_4^-(aq) + K^+(aq) + OH^-(aq) \longrightarrow K^+(aq) + ClO_4^-(aq) + H_2O(\ell)$

           $H^+(aq) + OH^-(aq) \longrightarrow H_2O(\ell)$

      (b) $2\,HBr(aq) + Ba(OH)_2(aq) \longrightarrow BaBr_2(aq) + 2\,H_2O(\ell)$

           $2\,H^+(aq) + 2\,Br^-(aq) + Ba^{2+}(aq) + 2\,OH^-(aq) \longrightarrow Ba^{2+}(aq) + 2\,Br^-(aq) + 2\,H_2O(\ell)$

           $H^+(aq) + OH^-(aq) \longrightarrow H_2O(\ell)$

      (c) $2\,HI(aq) + Ca(OH)_2(aq) \longrightarrow CaI_2(aq) + 2\,H_2O(\ell)$

           $2\,H^+(aq) + 2\,I^-(aq) + Ca^{2+}(aq) + 2\,OH^-(aq) \longrightarrow Ca^{2+}(aq) + 2\,I^-(aq) + 2\,H_2O(\ell)$

           $H^+(aq) + OH^-(aq) \longrightarrow H_2O(\ell)$

14.11   (a) $AgNO_3(aq) + NH_4Cl(aq) \longrightarrow AgCl(s) + NH_4NO_3(aq)$

           $Ag^+(aq) + NO_3^-(aq) + NH_4^+(aq) + Cl^-(aq) \longrightarrow AgCl(s) + NH_4^+(aq) + NO_3^-(aq)$

           $Ag^+(aq) + Cl^-(aq) \longrightarrow AgCl(s)$

      (b) $2\,AgNO_3(aq) + Na_2CO_3(aq) \longrightarrow Ag_2CO_3(s) + 2\,NaNO_3(aq)$

           $2\,Ag^+(aq) + 2\,NO_3^-(aq) + 2\,Na^+(aq) + CO_3^{2-}(aq) \longrightarrow$

                              $Ag_2CO_3(s) + 2\,Na^+(aq) + 2\,NO_3^-(aq)$

           $2\,Ag^+(aq) + CO_3^{2-}(aq) \longrightarrow Ag_2CO_3(s)$

      (c) $Sr(NO_3)_2(aq) + K_2SO_4(aq) \longrightarrow SrSO_4(s) + 2\,KNO_3(aq)$

           $Sr^{2+}(aq) + 2\,NO_3^-(aq) + 2\,K^+(aq) + SO_4^{2-}(aq) \longrightarrow SrSO_4(s) + 2\,K^+(aq) + 2\,NO_3^-(aq)$

           $Sr^{2+}(aq) + SO_4^{2-}(aq) \longrightarrow SrSO_4(s)$

14.12   (a) $K_2CO_3(aq) + 2\,HClO_4(aq) \longrightarrow 2\,KClO_4(aq) + H_2O(\ell) + CO_2(g)$

           $2\,K^+(aq) + CO_3^{2-}(aq) + 2\,H^+(aq) + 2\,ClO_4^-(aq) \longrightarrow$

                                $2\,K^+(aq) + 2\,ClO_4^-(aq) + H_2O(\ell) + CO_2(g)$

           $2\,H^+(aq) + CO_3^{2-}(aq) \longrightarrow H_2O(\ell) + CO_2(g)$

      (b) $Mg(s) + 2\,HNO_3(aq) \longrightarrow Mg(NO_3)_2(aq) + H_2(g)$

           $Mg(s) + 2\,H^+(aq) + 2\,NO_3^-(aq) \longrightarrow Mg^{2+}(aq) + 2\,NO_3^-(aq) + H_2(g)$

           $Mg(s) + 2\,H^+(aq) \longrightarrow Mg^{2+}(aq) + H_2(g)$

14.12 (continued)

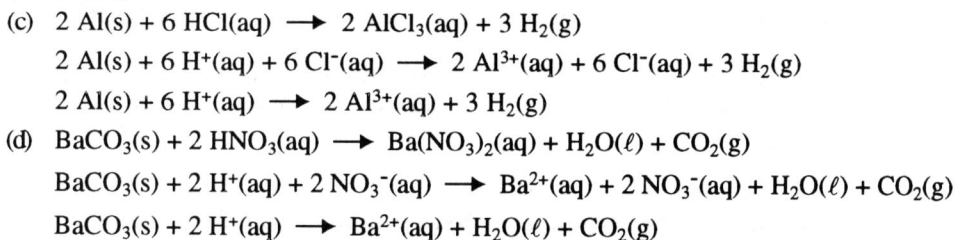

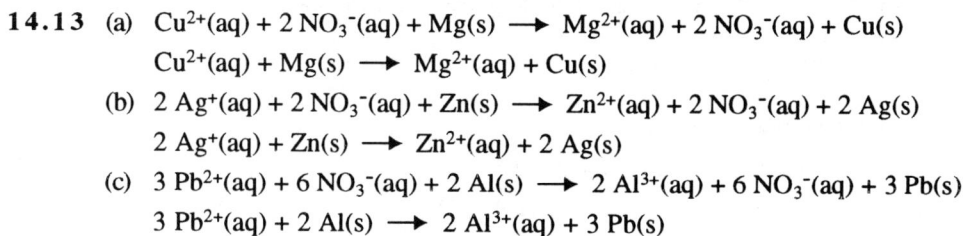

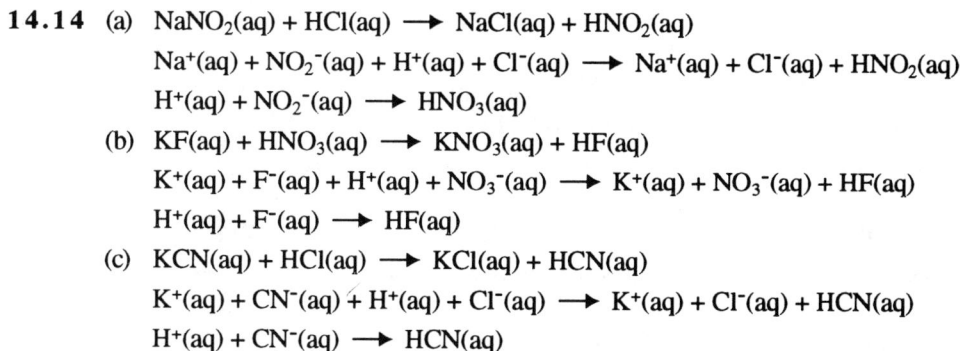

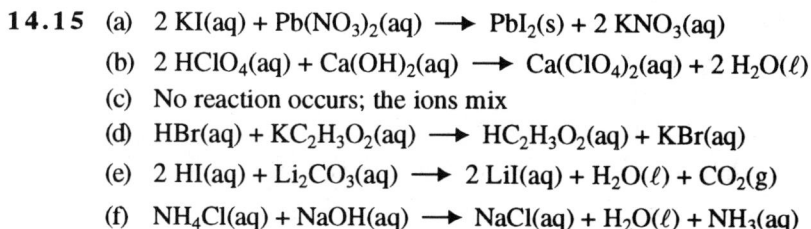

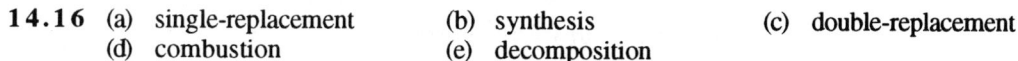

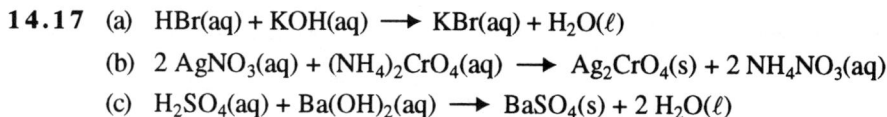

(c) $2 \text{ Al(s)} + 6 \text{ HCl(aq)} \longrightarrow 2 \text{ AlCl}_3(aq) + 3 \text{ H}_2(g)$

$2 \text{ Al(s)} + 6 \text{ H}^+(aq) + 6 \text{ Cl}^-(aq) \longrightarrow 2 \text{ Al}^{3+}(aq) + 6 \text{ Cl}^-(aq) + 3 \text{ H}_2(g)$

$2 \text{ Al(s)} + 6 \text{ H}^+(aq) \longrightarrow 2 \text{ Al}^{3+}(aq) + 3 \text{ H}_2(g)$

(d) $\text{BaCO}_3(s) + 2 \text{ HNO}_3(aq) \longrightarrow \text{Ba(NO}_3)_2(aq) + \text{H}_2\text{O}(\ell) + \text{CO}_2(g)$

$\text{BaCO}_3(s) + 2 \text{ H}^+(aq) + 2 \text{ NO}_3^-(aq) \longrightarrow \text{Ba}^{2+}(aq) + 2 \text{ NO}_3^-(aq) + \text{H}_2\text{O}(\ell) + \text{CO}_2(g)$

$\text{BaCO}_3(s) + 2 \text{ H}^+(aq) \longrightarrow \text{Ba}^{2+}(aq) + \text{H}_2\text{O}(\ell) + \text{CO}_2(g)$

14.13 (a) $\text{Cu}^{2+}(aq) + 2 \text{ NO}_3^-(aq) + \text{Mg(s)} \longrightarrow \text{Mg}^{2+}(aq) + 2 \text{ NO}_3^-(aq) + \text{Cu(s)}$

$\text{Cu}^{2+}(aq) + \text{Mg(s)} \longrightarrow \text{Mg}^{2+}(aq) + \text{Cu(s)}$

(b) $2 \text{ Ag}^+(aq) + 2 \text{ NO}_3^-(aq) + \text{Zn(s)} \longrightarrow \text{Zn}^{2+}(aq) + 2 \text{ NO}_3^-(aq) + 2 \text{ Ag(s)}$

$2 \text{ Ag}^+(aq) + \text{Zn(s)} \longrightarrow \text{Zn}^{2+}(aq) + 2 \text{ Ag(s)}$

(c) $3 \text{ Pb}^{2+}(aq) + 6 \text{ NO}_3^-(aq) + 2 \text{ Al(s)} \longrightarrow 2 \text{ Al}^{3+}(aq) + 6 \text{ NO}_3^-(aq) + 3 \text{ Pb(s)}$

$3 \text{ Pb}^{2+}(aq) + 2 \text{ Al(s)} \longrightarrow 2 \text{ Al}^{3+}(aq) + 3 \text{ Pb(s)}$

14.14 (a) $\text{NaNO}_2(aq) + \text{HCl(aq)} \longrightarrow \text{NaCl(aq)} + \text{HNO}_2(aq)$

$\text{Na}^+(aq) + \text{NO}_2^-(aq) + \text{H}^+(aq) + \text{Cl}^-(aq) \longrightarrow \text{Na}^+(aq) + \text{Cl}^-(aq) + \text{HNO}_2(aq)$

$\text{H}^+(aq) + \text{NO}_2^-(aq) \longrightarrow \text{HNO}_3(aq)$

(b) $\text{KF(aq)} + \text{HNO}_3(aq) \longrightarrow \text{KNO}_3(aq) + \text{HF(aq)}$

$\text{K}^+(aq) + \text{F}^-(aq) + \text{H}^+(aq) + \text{NO}_3^-(aq) \longrightarrow \text{K}^+(aq) + \text{NO}_3^-(aq) + \text{HF(aq)}$

$\text{H}^+(aq) + \text{F}^-(aq) \longrightarrow \text{HF(aq)}$

(c) $\text{KCN(aq)} + \text{HCl(aq)} \longrightarrow \text{KCl(aq)} + \text{HCN(aq)}$

$\text{K}^+(aq) + \text{CN}^-(aq) + \text{H}^+(aq) + \text{Cl}^-(aq) \longrightarrow \text{K}^+(aq) + \text{Cl}^-(aq) + \text{HCN(aq)}$

$\text{H}^+(aq) + \text{CN}^-(aq) \longrightarrow \text{HCN(aq)}$

14.15 (a) $2 \text{ KI(aq)} + \text{Pb(NO}_3)_2(aq) \longrightarrow \text{PbI}_2(s) + 2 \text{ KNO}_3(aq)$

(b) $2 \text{ HClO}_4(aq) + \text{Ca(OH)}_2(aq) \longrightarrow \text{Ca(ClO}_4)_2(aq) + 2 \text{ H}_2\text{O}(\ell)$

(c) No reaction occurs; the ions mix

(d) $\text{HBr(aq)} + \text{KC}_2\text{H}_3\text{O}_2(aq) \longrightarrow \text{HC}_2\text{H}_3\text{O}_2(aq) + \text{KBr(aq)}$

(e) $2 \text{ HI(aq)} + \text{Li}_2\text{CO}_3(aq) \longrightarrow 2 \text{ LiI(aq)} + \text{H}_2\text{O}(\ell) + \text{CO}_2(g)$

(f) $\text{NH}_4\text{Cl(aq)} + \text{NaOH(aq)} \longrightarrow \text{NaCl(aq)} + \text{H}_2\text{O}(\ell) + \text{NH}_3(aq)$

14.16 (a) single-replacement   (b) synthesis   (c) double-replacement

(d) combustion   (e) decomposition

14.17 (a) $\text{HBr(aq)} + \text{KOH(aq)} \longrightarrow \text{KBr(aq)} + \text{H}_2\text{O}(\ell)$

(b) $2 \text{ AgNO}_3(aq) + (\text{NH}_4)_2\text{CrO}_4(aq) \longrightarrow \text{Ag}_2\text{CrO}_4(s) + 2 \text{ NH}_4\text{NO}_3(aq)$

(c) $\text{H}_2\text{SO}_4(aq) + \text{Ba(OH)}_2(aq) \longrightarrow \text{BaSO}_4(s) + 2 \text{ H}_2\text{O}(\ell)$

14.18 (a) No. It is a nonelectrolyte.

(b) The sodium sulfate permits the aqueous solution to conduct electricity.

14.19 A strong acid completely dissociates. A weak acid does not. Dissolve each in water and test the conductivity of each solution. The strong acid will conduct strongly. The weak acid will conduct weakly.

14.20 (a) A proton refers to a hydrogen ion.

(b) A diprotic acid has two ionizable hydrogens in its formula.

(c) A tripotic acid has three ionizable hydrogens in its formula.

(d) Alkaline means basic.

(e) A hydronium ion, H_3O^+, is the form of the hydrogen ion in aqueous solution.

14.21 (a) $HI + NaOH \longrightarrow NaI + H_2O$

(b) $H_2SO_4 + 2\,KOH \longrightarrow K_2SO_4 + 2\,H_2O$

(c) $2\,HNO_3 + Ca(OH)_2 \longrightarrow Ca(NO_3)_2 + 2\,H_2O$

(d) $H_3PO_4 + 3\,NaOH \longrightarrow Na_3PO_4 + 3\,H_2O$

14.22 (a) $HBr + KOH \longrightarrow KBr + H_2O$

(b) $2\,HNO_3 + Ba(OH)_2 \longrightarrow Ba(NO_3)_2 + 2\,H_2O$

(c) $H_2SO_4 + Mg(OH)_2 \longrightarrow MgSO_4 + 2\,H_2O$

(d) $H_3PO_4 + 3\,KOH \longrightarrow K_3PO_4 + 3\,H_2O$

14.23 (a) $K_2CO_3 + 2\,HBr \longrightarrow 2\,KBr + H_2O + CO_2$

(b) $(NH_4)_2CO_3 + 2\,HCl \longrightarrow 2\,NH_4Cl + H_2O + CO_2$

14.24 (a) $Ca^{2+}(aq)$ and $I^-(aq)$ in a $1:2$ ratio.

(b) $NH_4^+(aq)$ and CrO_4^{2-} in a $2:1$ ratio.

(c) $Mg^{2+}(aq)$ and $NO_3^-(aq)$ in a $1:2$ ratio.

(d) $Al^{3+}(aq)$ and $SO_4^{2-}(aq)$ in a $2:3$ ratio.

14.25 (a) $HI(aq) + NaOH(aq) \longrightarrow NaI(aq) + H_2O(\ell)$

$H^+(aq) + I^-(aq) + Na^+(aq) + OH^-(aq) \longrightarrow Na^+(aq) + I^-(aq) + H_2O(\ell)$

$H^+(aq) + OH^-(aq) \longrightarrow H_2O(\ell)$

(b) $2\,HClO_4(aq) + Ca(OH)_2(aq) \longrightarrow Ca(ClO_4)_2(aq) + 2\,H_2O(\ell)$

$2\,H^+(aq) + 2\,ClO_4^-(aq) + Ca^{2+}(aq) + 2\,OH^-(aq) \longrightarrow$

$$Ca^{2+}(aq) + 2\,ClO_4^-(aq) + 2\,H_2O(\ell)$$

$H^+(aq) + OH^-(aq) \longrightarrow H_2O(\ell)$

14.26 (a) Sodium ions, $Na^+(aq)$, and chloride ions, $Cl^-(aq)$.

(b) Silver ions, $Ag^+(aq)$, and nitrate ions, $NO_3^-(aq)$.

(c) Silver ions combine with chloride ions to form a precipitate, which is no longer in solution.

(d) Sodium ions and nitrate ions are spectator ions. These ions remain unreacted in the solution.

14.27 (a) $Pb(NO_3)_2(aq) + Na_2SO_4(aq) \longrightarrow PbSO_4(s) + 2\,NaNO_3(aq)$

$Pb^{2+}(aq) + 2\,NO_3^-(aq) + 2\,Na^+(aq) + SO_4^{2-}(aq) \longrightarrow$

$$PbSO_4(s) + 2\,Na^+(aq) + 2\,NO_3^-(aq)$$

$Pb^{2+}(aq) + SO_4^{2-}(aq) \longrightarrow PbSO_4(s)$

(b) $2\,AgNO_3(aq) + MgBr_2(aq) \longrightarrow 2\,AgBr(s) + Mg(NO_3)_2(aq)$

$2\,Ag^+(aq) + 2\,NO_3^-(aq) + Mg^{2+}(aq) + 2\,Br^-(aq) \longrightarrow$

$$2\,AgBr(s) + Mg^{2+}(aq) + 2\,NO_3^-(aq)$$

$Ag^+(aq) + Br^-(aq) \longrightarrow AgBr(s)$

(c) $MgCl_2(aq) + Na_2CO_3(aq) \longrightarrow MgCO_3(s) + 2\,NaCl(aq)$

$Mg^{2+}(aq) + 2\,Cl^-(aq) + 2\,Na^+(aq) + CO_3^{2-}(aq) \longrightarrow$

$$MgCO_3(s) + 2\,Na^+(aq) + 2\,Cl^-(aq)$$

$Mg^{2+}(aq) + CO_3^{2-}(aq) \longrightarrow MgCO_3(s)$

14.28 (a) $2\,H^+(aq) + 2\,NO_3^-(aq) + 2\,NH_4^+(aq) + CO_3^{2-}(aq) \longrightarrow$

$$2\,NH_4^+(aq) + 2\,NO_3^-(aq) + H_2O(\ell) + CO_2(g)$$

$2\,H^+(aq) + CO_3^{2-}(aq) \longrightarrow H_2O(\ell) + CO_2(g)$

(b) $Zn(s) + 2\,H^+(aq) + 2\,Cl^-(aq) \longrightarrow Zn^{2+}(aq) + 2\,Cl^-(aq) + H_2(g)$

$Zn(s) + 2\,H^+(aq) \longrightarrow Zn^{2+}(aq) + H_2(g)$

(c) $CaCO_3(s) + 2\,H^+(aq) + 2\,I^-(aq) \longrightarrow Ca^{2+}(aq) + 2\,I^-(aq) + H_2O(\ell) + CO_2(g)$

$CaCO_3(s) + 2\,H^+(aq) \longrightarrow Ca^{2+}(aq) + H_2O(\ell) + CO_2(g)$

(d) $2\,Al(s) + 3\,Cu^{2+}(aq) + 6\,NO_3^-(aq) \longrightarrow 2\,Al^{3+}(aq) + 6\,NO_3^-(aq) + 3\,Cu(s)$

$2\,Al(s) + 3\,Cu^{2+}(aq) \longrightarrow 2\,Al^{3+}(aq) + 3\,Cu(s)$

(e) $2\,Na(s) + 2\,H_2O(\ell) \longrightarrow 2\,Na^+(aq) + 2\,OH^-(aq) + H_2(g)$

There are no spectators; the net ionic equation is the same as the total ionic equation.

(f) $3\,Cu(s) + 8\,H^+(aq) + 8\,NO_3^-(aq) \longrightarrow$

$$3\,Cu^{2+}(aq) + 6\,NO_3^-(aq) + 2\,NO(g) + 4\,H_2O(\ell)$$

$3\,Cu(s) + 8\,H^+(aq) + 2\,NO_3^-(aq) \longrightarrow 3\,Cu^{2+}(aq) + 2\,NO(g) + 4\,H_2O(\ell)$

(g) $H^+(aq) + I^-(aq) + K^+(aq) + NO_2^-(aq) \longrightarrow K^+(aq) + I^-(aq) + HNO_2(aq)$

$H^+(aq) + NO_2^-(aq) \longrightarrow HNO_2(aq)$

(h) $HCN(aq) + Na^+(aq) + OH^-(aq) \longrightarrow Na^+(aq) + CN^-(aq) + H_2O(\ell)$

$HCN(aq) + OH^-(aq) \longrightarrow CN^-(aq) + H_2O(\ell)$

14.29 total molecular equation:

$$CaCO_3(s) + 2\ HC_2H_3O_2(aq) \longrightarrow Ca(C_2H_3O_2)_2(aq) + H_2O(\ell) + CO_2(g)$$

total ionic equation and net ionic equatoion:

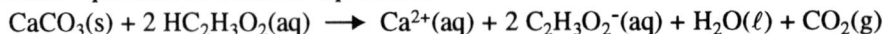

$$CaCO_3(s) + 2\ HC_2H_3O_2(aq) \longrightarrow Ca^{2+}(aq) + 2\ C_2H_3O_2^-(aq) + H_2O(\ell) + CO_2(g)$$

14.30 (a) $BaCl_2(aq) + K_2SO_4(aq) \longrightarrow BaSO_4(s) + 2\ KCl(aq)$

(b) no reaction; mixture of ions

(c) $HI(aq) + KOH(aq) \longrightarrow KI(aq) + H_2O(\ell)$

(d) $HCl(aq) + NaCN(aq) \longrightarrow HCN(aq) + NaCl(aq)$

(e) $H_2SO_4(aq) + Na_2CO_3(aq) \longrightarrow Na_2SO_4(aq) + H_2O(\ell) + CO_2(g)$

14.31 As distilled water is added, the concentration of ions will decrease, causing the bulb to dim.

14.32 $H^+(aq) + HCO_3^-(aq) \longrightarrow CO_2(g) + H_2O(\ell)$

$HCO_3^-(aq) + OH^-(aq) \longrightarrow CO_3^{2-}(aq) + H_2O(\ell)$

14.33 Sodium hydrogen carbonate is capable of neutralizing acids by forming a salt, water, and carbon dioxide upon reaction with an acid. In the case of hydrochloric acid, the following neutralization reaction would take place:

$$HCl + NaHCO_3 \longrightarrow NaCl + H_2O + CO_2$$

$$125\ mL\ soln\left(\frac{1\ L\ soln}{1000\ mL\ soln}\right)\left(\frac{0.100\ mol\ HCl}{1\ L\ soln}\right) = 0.0125\ mol\ HCl$$

$$0.0125\ mol\ HCl\left(\frac{1\ mol\ NaHCO_3}{1\ mol\ HCl}\right)\left(\frac{84.0\ g\ NaHCO_3}{1\ mol\ NaHCO_3}\right) = 1.05\ g\ NaHCO_3$$

14.34 (a) The presence of ions from the sulfuric acid allows the solution to conduct electricity.

(b) The light goes out at the neutralization point.

(c) The hydroxide ions combine with hydrogen ions to form water, a nonelectrolyte. The barium ions combine with sulfate ions to form a precipitate, thereby removing each of these from solution. At the neutralization point there is not a large enough concentration of ions left in solution to conduct electricity well.

(d) The further addition of barium hydroxide reintroduces ions into solution, thereby permitting the flow of electricity.

14.35 (a) $H_2C_2O_4 + NaOH \longrightarrow NaHC_2O_4 + H_2O$

(b) $NaHC_2O_4 + NaOH \longrightarrow Na_2C_2O_4 + H_2O$

(c) $H_3PO_4 + KOH \longrightarrow KH_2PO_4 + H_2O$

14.35 (continued)

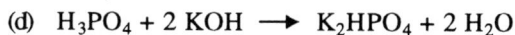

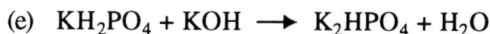

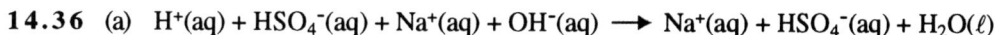

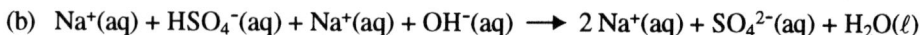

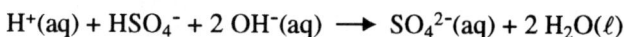

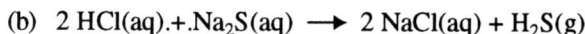

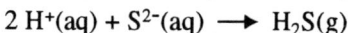

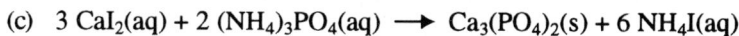

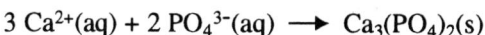

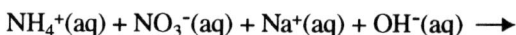

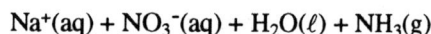

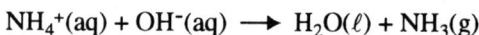

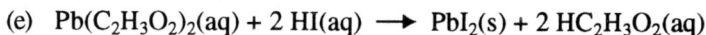

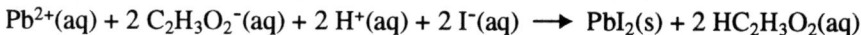

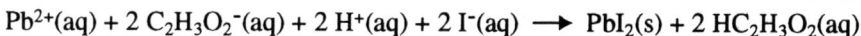

    (d) $H_3PO_4 + 2\ KOH \longrightarrow K_2HPO_4 + 2\ H_2O$

    (e) $KH_2PO_4 + KOH \longrightarrow K_2HPO_4 + H_2O$

14.36 (a) $H^+(aq) + HSO_4^-(aq) + Na^+(aq) + OH^-(aq) \longrightarrow Na^+(aq) + HSO_4^-(aq) + H_2O(\ell)$

        $H^+(aq) + OH^-(aq) \longrightarrow H_2O$

    (b) $Na^+(aq) + HSO_4^-(aq) + Na^+(aq) + OH^-(aq) \longrightarrow 2\ Na^+(aq) + SO_4^{2-}(aq) + H_2O(\ell)$

        $HSO_4^-(aq) + OH^-(aq) \longrightarrow SO_4^{2-}(aq) + H_2O(\ell)$

    (c) $H^+(aq) + HSO_4^-(aq) + 2\ Na^+(aq) + 2\ OH^-(aq) \longrightarrow 2\ Na^+(aq) + SO_4^{2-}(aq) + 2\ H_2O(\ell)$

        $H^+(aq) + HSO_4^- + 2\ OH^-(aq) \longrightarrow SO_4^{2-}(aq) + 2\ H_2O(\ell)$

14.37 (a) $2\ Al(s) + 6\ HBr(aq) + 2\ AlBr_3(aq) + 3\ H_2(g)$

        $2\ Al(s) + 6\ H^+(aq) + 6\ Br^-(aq) \longrightarrow 2\ Al^{3+}(aq) + 6\ Br^-(aq) + 3\ H_2(g)$

        $2\ Al(s) + 6\ H^+(aq) \longrightarrow 2\ Al^{3+}(aq) + 3\ H_2(g)$

    (b) $2\ HCl(aq).+.Na_2S(aq) \longrightarrow 2\ NaCl(aq) + H_2S(g)$

        $2\ H^+(aq) + 2\ Cl^-(aq) + 2\ Na^+(aq) + S^{2-}(aq) \longrightarrow 2\ Na^+(aq) + 2\ Cl^-(aq) + H_2S(g)$

        $2\ H^+(aq) + S^{2-}(aq) \longrightarrow H_2S(g)$

    (c) $3\ CaI_2(aq) + 2\ (NH_4)_3PO_4(aq) \longrightarrow Ca_3(PO_4)_2(s) + 6\ NH_4I(aq)$

        $3\ Ca^{2+}(aq) + 6\ I^-(aq) + 6\ NH_4^+(aq) + 2\ PO_4^{3-}(aq) \longrightarrow$

                                    $Ca_3(PO_4)_2(s) + 6\ NH_4^+(aq) + 6\ I^-(aq)$

        $3\ Ca^{2+}(aq) + 2\ PO_4^{3-}(aq) \longrightarrow Ca_3(PO_4)_2(s)$

    (d) $NH_4NO_3(aq) + NaOH(aq) \longrightarrow NaNO_3(aq) + H_2O(\ell) + NH_3(g)$

        $NH_4^+(aq) + NO_3^-(aq) + Na^+(aq) + OH^-(aq) \longrightarrow$

                                  $Na^+(aq) + NO_3^-(aq) + H_2O(\ell) + NH_3(g)$

        $NH_4^+(aq) + OH^-(aq) \longrightarrow H_2O(\ell) + NH_3(g)$

    (e) $Pb(C_2H_3O_2)_2(aq) + 2\ HI(aq) \longrightarrow PbI_2(s) + 2\ HC_2H_3O_2(aq)$

        $Pb^{2+}(aq) + 2\ C_2H_3O_2^-(aq) + 2\ H^+(aq) + 2\ I^-(aq) \longrightarrow PbI_2(s) + 2\ HC_2H_3O_2(aq)$

        $Pb^{2+}(aq) + 2\ C_2H_3O_2^-(aq) + 2\ H^+(aq) + 2\ I^-(aq) \longrightarrow PbI_2(s) + 2\ HC_2H_3O_2(aq)$

14.38 $HCl(g) + NH_3(g) \longrightarrow NH_4Cl(s)$

CHAPTER 15

Working with Acids and Bases

15.1 (a) $[OH^-] = 0.30$ M (b) $[H^+] = 0.50$ M
(c) $[OH^-] = 0.65$ M (d) $[H^+] = 0.73$ M

15.2 (a) $[OH^-] = \dfrac{1.0 \times 10^{-14}}{1.0 \times 10^{-10}} = 1.0 \times 10^{-4}$ (b) $[H^+] = \dfrac{1.0 \times 10^{-14}}{1.0 \times 10^{-6}} = 1.0 \times 10^{-8}$

(c) $[OH^-] = \dfrac{1.0 \times 10^{-14}}{2.0 \times 10^{-5}} = 5.0 \times 10^{-10}$ (d) $[H^+] = \dfrac{1.0 \times 10^{-14}}{6.5 \times 10^{-4}} = 1.5 \times 10^{-11}$

15.3 (a) pOH = 7.00; neutral   (b) pOH = 6.50; weakly basic
(c) pOH = 8.20; weakly acidic   (d) pOH = 2.30; strongly basic
(e) pOH = 11.4; strongly acidic

15.4 (a) $pH = -\log[1.0 \times 10^{-4}] = -(-4.00) = 4.00$; pOH = 10.00

(b) $pOH = -\log[1.0 \times 10^{-4}] = -(-4.00) = 4.00$; pH = 10.00

(c) $pH = -\log[1.0 \times 10^{-2}] = -(-2.00) = 2.00$; pOH = 12.00

(d) $pOH = -\log[1.0 \times 10^{-5}] = -(-5.00) = 5.00$; pH = 9.00

(e) $pH = -\log[1.0 \times 10^{0}] = -(0.00) = 0.00$; pOH = 14.00

(f) $pH = -\log[6.8 \times 10^{-2}] = -(-1.17) = 1.17$; pOH = 12.83

(g) $pH = -\log[3.1 \times 10^{-4}] = -(-3.51) = 3.51$; pOH = 10.49

(h) $pH = -\log[5.8 \times 10^{-5}] = -(-4.24) = 4.24$; pOH = 9.76

(i) $pOH = -\log[2.2 \times 10^{-2}] = -(-1.66) = 1.66$; pH = 12.34

(j) $pOH = -\log[7.4 \times 10^{-3}] = -(-2.13) = 2.13$; pH = 11.87

15.5 (a) $0.63 \text{ g HNO}_3 \left(\dfrac{1 \text{ mol HNO}_3}{63.0 \text{ g HNO}_3} \right) = 0.010 \text{ mol HNO}_3$

$\text{molarity} = \dfrac{0.010 \text{ mol HNO}_3}{1000 \text{ L soln}} = 1.0 \times 10^{-5} \text{ M HNO}_3$

$[H^+] = 1.0 \times 10^{-5}$ M

$pH = -\log(1.0 \times 10^{-5}) = -(-5.00) = 5.00$

15.5 (continued)

(b) $4.0 \text{ g NaOH} \left(\dfrac{1 \text{ mol NaOH}}{40.0 \text{ g NaOH}} \right) = 0.10 \text{ mol NaOH}$

  molarity $= \dfrac{0.10 \text{ mol NaOH}}{100 \text{ L soln}} = 1.0 \times 10^{-3} \text{ M NaOH}$

  $[OH^-] = 1.0 \times 10^{-3} \text{ M}$
  $pOH = -\log(1.0 \times 10^{-3}) = -(-3.00) = 3.00$
  $pH = 11.00$

(c) $0.080 \text{ g NaOH} \left(\dfrac{1 \text{ mol NaOH}}{40.0 \text{ g NaOH}} \right) = 0.0020 \text{ mol NaOH}$

  molarity $= \dfrac{0.0020 \text{ mol NaOH}}{0.200 \text{ L soln}} = 1.0 \times 10^{-2} \text{ M NaOH}$

  $[OH^-] = 1.0 \times 10^{-2} \text{ M}$
  $pOH = -\log(1.0 \times 10^{-2}) = -(-2.00) = 2.00$
  $pH = 12.00$

(d) $73 \text{ g HCl} \left(\dfrac{1 \text{ mol HCl}}{36.5 \text{ g HCl}} \right) = 2.0 \text{ mol HCl}$

  molarity $= \dfrac{2.0 \text{ mol HCl}}{20.0 \text{ L soln}} = 1.0 \times 10^{-1} \text{ M HCl}$

  $[H^+] = 1.0 \times 10^{-1} \text{ M}$
  $pH = -\log(1.0 \times 10^{-1}) = -(-1.00) = 1.00$

(e) $0.051 \text{ g HCl} \left(\dfrac{1 \text{ mol HCl}}{36.5 \text{ g HCl}} \right) = 1.4 \times 10^{-3} \text{ mol HCl}$

  molarity $= \dfrac{1.4 \times 10^{-3} \text{ mol HCl}}{15 \text{ L soln}} = 9.3 \times 10^{-5} \text{ M HCl}$

  $[H^+] = 9.3 \times 10^{-5} \text{ M}$
  $pH = -\log(9.3 \times 10^{-5}) = -(-4.03) = 4.03$

(f) $0.10 \text{ g NaOH} \left(\dfrac{1 \text{ mol NaOH}}{40.0 \text{ g NaOH}} \right) = 2.5 \times 10^{-3} \text{ mol NaOH}$

  molarity $= \dfrac{2.5 \times 10^{-3} \text{ mol NaOH}}{5.0 \text{ L soln}} = 5.0 \times 10^{-4} \text{ M NaOH}$

  $[OH^-] = 5.0 \times 10^{-4} \text{ M}$
  $pOH = -\log(5.0 \times 10^{-4}) = -(-3.30) = 3.30$
  $pH = 10.70$

15.5 (continued)

(g) $0.084 \text{ g KOH}\left(\dfrac{1 \text{ mol KOH}}{56.1 \text{ g KOH}}\right) = 1.5 \times 10^{-3} \text{ mol KOH}$

$\text{molarity} = \dfrac{1.5 \times 10^{-3} \text{ mol KOH}}{0.250 \text{ L soln}} = 6.0 \times 10^{-3} \text{ M KOH}$

$[OH^-] = 6.0 \times 10^{-3} \text{ M}$
$pOH = -\log(6.0 \times 10^{-3}) = -(-2.22) = 2.22$
$pH = 11.78$

(h) $0.59 \text{ g HBr}\left(\dfrac{1 \text{ mol HBr}}{80.9 \text{ g HBr}}\right) = 7.3 \times 10^{-3} \text{ mol HBr}$

$\text{molarity} = \dfrac{7.3 \times 10^{-3} \text{ mol HBr}}{4.0 \text{ L soln}} = 1.8 \times 10^{-3} \text{ M HBr}$

$[H^+] = 1.8 \times 10^{-3} \text{ M}$
$pH = -\log(1.8 \times 10^{-3}) = -(-2.74) = 2.74$

15.6 (a) $[H^+] = 1.0 \times 10^{-4} \text{ M}$ (b) $[H^+] = 1.0 \times 10^{-11} \text{ M}$

(c) $pH = 12.00; [H^+] = 1.0 \times 10^{-12} \text{ M}$ (d) $[H^+] = 3.6 \times 10^{-9} \text{ M}$

(e) $[H^+] = 1.3 \times 10^{-3} \text{ M}$ (f) $pH = 3.67; [H^+] = 2.1 \times 10^{-4} \text{ M}$

15.7 (a)  $ClO^- + H_2O \rightleftharpoons HClO + OH^-$ (b)  $NO_2^- + H_2O \rightleftharpoons HNO_2 + OH^-$

(c) 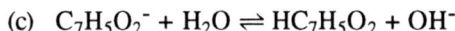 $C_7H_5O_2^- + H_2O \rightleftharpoons HC_7H_5O_2 + OH^-$

15.8 (a) Br^- (b) pH_4^+ (c) $HC_2O_4^-$ (d) HS^- (e) S^{2-} (f) $HC_2O_4^-$ (g) $H_2C_2O_4$

15.9 (a) H_2SO_4 (b) SO_4^{2-}

15.10

	acid	*base*
(a)	HCN	OH^-
(b)	H_2SO_4	NH_3
(c)	HNO_3	H_2O
(d)	H_2O	F^-
(e)	HSO_4^-	HPO_4^{2-}

15.11

	acid	*base*
(a)	H^+	pH_3
(b)	Cu^+	NH_3
(c)	BF_3	NH_3
(d)	$FeCl_3$	Cl^-

15.12 (a) no   (b) yes

15.13 (a) Will work as a buffer
        (b) Will not work as a buffer
        (c) Will work as a buffer
        (d) Will not work as a buffer

15.14 (a) 3 (b) 1 (c) 3 (d) 1 (e) 2 (f) 2

15.15 (a) $0.250 \text{ mol HCl} \left(\dfrac{1 \text{ equiv HCl}}{1 \text{ mol HCl}} \right) = 0.250 \text{ equiv HCl}$

     (b) $0.550 \text{ mol Ba(OH)}_2 \left(\dfrac{2 \text{ equiv Ba(OH)}_2}{1 \text{ mol Ba(OH)}_2} \right) = 1.10 \text{ equiv Ba(OH)}_2$

     (c) $2.50 \text{ mol H}_3\text{PO}_4 \left(\dfrac{3 \text{ equiv H}_3\text{PO}_4}{1 \text{ mol H}_3\text{PO}_4} \right) = 7.50 \text{ equiv H}_3\text{PO}_4$

     (d) $0.0140 \text{ mol H}_2\text{SO}_4 \left(\dfrac{2 \text{ equiv H}_2\text{SO}_4}{1 \text{ mol H}_2\text{SO}_4} \right) = 0.0280 \text{ equiv H}_2\text{SO}_4$

     (e) $0.630 \text{ mol Al(OH)}_3 \left(\dfrac{3 \text{ equiv Al(OH)}_3}{1 \text{ mol Al(OH)}_3} \right) = 1.89 \text{ equiv Al(OH)}_3$

15.16 (a) $0.150 \text{ equiv H}_3\text{PO}_4 \left(\dfrac{1 \text{ mol H}_3\text{PO}_4}{3 \text{ equiv H}_3\text{PO}_4} \right) = 0.0500 \text{ mol H}_3\text{PO}_4$

     (b) $0.650 \text{ equiv HBr} \left(\dfrac{1 \text{ mol HBr}}{1 \text{ equiv HBr}} \right) = 0.650 \text{ mol HBr}$

     (c) $1.55 \text{ equiv Ca(OH)}_2 \left(\dfrac{1 \text{ mol Ca(OH)}_2}{2 \text{ equiv Ca(OH)}_2} \right) = 0.775 \text{ mol Ca(OH)}_2$

     (d) $3.25 \text{ equiv KOH} \left(\dfrac{1 \text{ mol KOH}}{1 \text{ equiv KOH}} \right) = 3.25 \text{ mol KOH}$

     (e) $0.0750 \text{ equiv H}_2\text{C}_2\text{O}_4 \left(\dfrac{1 \text{ mol H}_2\text{C}_2\text{O}_4}{2 \text{ equiv H}_2\text{C}_2\text{O}_4} \right) = 0.0375 \text{ mol H}_2\text{C}_2\text{O}_4$

15.17 (a) $3.65 \text{ g HCl} \left(\dfrac{1 \text{ mol HCl}}{36.5 \text{ g HCl}} \right)\left(\dfrac{1 \text{ equiv HCl}}{1 \text{ mol HCl}} \right) = 0.100 \text{ equiv HCl}$

     (b) $0.0200 \text{ g NaOH} \left(\dfrac{1 \text{ mol NaOH}}{40.0 \text{ g NaOH}} \right)\left(\dfrac{1 \text{ equiv NaOH}}{1 \text{ mol NaOH}} \right) = 5.00 \times 10^{-4} \text{ equiv NaOH}$

15.17 (continued)

(c) $7.41 \text{ g Ca(OH)}_2 \left(\dfrac{1 \text{ mol Ca(OH)}_2}{74.1 \text{ g Ca(OH)}_2} \right) \left(\dfrac{2 \text{ equiv Ca(OH)}_2}{1 \text{ mol Ca(OH)}_2} \right) = 0.200 \text{ equiv Ca(OH)}_2$

(d) $0.162 \text{ g HBr} \left(\dfrac{1 \text{ mol HBr}}{80.9 \text{ g HBr}} \right) \left(\dfrac{1 \text{ equiv HBr}}{1 \text{ mol HBr}} \right) = 0.00200 \text{ equiv HBr}$

(e) $21.0 \text{ g H}_2\text{C}_2\text{O}_4 \left(\dfrac{1 \text{ mol H}_2\text{C}_2\text{O}_4}{90.0 \text{ g H}_2\text{C}_2\text{O}_4} \right) \left(\dfrac{2 \text{ equiv H}_2\text{C}_2\text{O}_4}{1 \text{ mol H}_2\text{C}_2\text{O}_4} \right) = 0.467 \text{ equiv H}_2\text{C}_2\text{O}_4$

(f) $2.94 \text{ g H}_2\text{SO}_4 \left(\dfrac{1 \text{ mol H}_2\text{SO}_4}{98.1 \text{ g H}_2\text{SO}_4} \right) \left(\dfrac{2 \text{ equiv H}_2\text{SO}_4}{1 \text{ mol H}_2\text{SO}_4} \right) = 0.0599 \text{ equiv H}_2\text{SO}_4$

(g) $0.392 \text{ g H}_3\text{PO}_4 \left(\dfrac{1 \text{ mol H}_3\text{PO}_4}{98.0 \text{ g H}_3\text{PO}_4} \right) \left(\dfrac{3 \text{ equiv H}_3\text{PO}_4}{1 \text{ mol H}_3\text{PO}_4} \right) = 0.0120 \text{ equiv H}_3\text{PO}_4$

15.18 (a) $1.00 \text{ g NaOH} \left(\dfrac{1 \text{ mol NaOH}}{40.0 \text{ g NaOH}} \right) \left(\dfrac{1 \text{ equiv NaOH}}{1 \text{ mol NaOH}} \right) = 0.0250 \text{ equiv NaOH}$

normality $= \dfrac{0.0250 \text{ equiv NaOH}}{0.0500 \text{ L soln}} = 0.500 \text{ N NaOH}$

(b) $0.500 \text{ g H}_2\text{SO}_4 \left(\dfrac{1 \text{ mol H}_2\text{SO}_4}{98.1 \text{ g H}_2\text{SO}_4} \right) \left(\dfrac{2 \text{ equiv H}_2\text{SO}_4}{1 \text{ mol H}_2\text{SO}_4} \right) = 0.0102 \text{ equiv H}_2\text{SO}_4$

normality $= \dfrac{0.0102 \text{ equiv H}_2\text{SO}_4}{0.250 \text{ L soln}} = 0.0408 \text{ N H}_2\text{SO}_4$

(c) $2.00 \text{ g H}_3\text{PO}_4 \left(\dfrac{1 \text{ mol H}_3\text{PO}_4}{98.0 \text{ g H}_3\text{PO}_4} \right) \left(\dfrac{3 \text{ equiv H}_3\text{PO}_4}{1 \text{ mol H}_3\text{PO}_4} \right) = 0.0612 \text{ equiv H}_3\text{PO}_4$

normality $= \dfrac{0.0612 \text{ equiv H}_3\text{PO}_4}{0.150 \text{ L soln}} = 0.408 \text{ N H}_3\text{PO}_4$

(d) $0.350 \text{ g Mg(OH)}_2 \left(\dfrac{1 \text{ mol Mg(OH)}_2}{58.3 \text{ g Mg(OH)}_2} \right) \left(\dfrac{2 \text{ equiv Mg(OH)}_2}{1 \text{ mol Mg(OH)}_2} \right)$

$= 0.0120 \text{ equiv Mg(OH)}_2$

normality $= \dfrac{0.0120 \text{ equiv Mg(OH)}_2}{2.50 \text{ L soln}} = 0.00480 \text{ N Mg(OH)}_2$

(e) $9.87 \text{ g HBr} \left(\dfrac{1 \text{ mol HBr}}{80.9 \text{ g HBr}} \right) \left(\dfrac{1 \text{ equiv HBr}}{1 \text{ mol HBr}} \right) = 0.122 \text{ equiv HBr}$

normality $= \dfrac{0.122 \text{ equiv HBr}}{1.80 \text{ L soln}} = 0.0678 \text{ N HBr}$

15.19 (a) $2.00 \text{ L soln}\left(\dfrac{0.300 \text{ equiv } H_2SO_4}{\text{L soln}}\right) = 0.600 \text{ equiv } H_2SO_4$

(b) $0.500 \text{ L soln}\left(\dfrac{1.20 \text{ equiv } HCl}{\text{L soln}}\right) = 0.600 \text{ equiv } HCl$

(c) $0.350 \text{ L soln}\left(\dfrac{0.750 \text{ equiv } NaOH}{\text{L soln}}\right) = 0.262 \text{ equiv } NaOH$

(d) $0.175 \text{ L soln}\left(\dfrac{0.180 \text{ equiv } H_3PO_4}{\text{L soln}}\right) = 0.0315 \text{ equiv } H_3PO_4$

(e) $0.0500 \text{ L soln}\left(\dfrac{0.125 \text{ equiv } H_2C_2O_4}{\text{L soln}}\right) = 0.00625 \text{ equiv } H_2C_2O_4$

15.20 $N_a = \dfrac{V_b \cdot N_b}{V_a} = \dfrac{(45.0 \text{ mL soln})(0.200 \text{ N})}{(55.0 \text{ mL soln})} = 0.164 \text{ N}$

15.21 $N_b = \dfrac{V_a \cdot N_a}{V_b} = \dfrac{(25.0 \text{ mL soln})(0.450 \text{ N})}{(35.0 \text{ mL soln})} = 0.321 \text{ N}$

15.22 $N_b = \dfrac{V_a \cdot N_a}{V_b} = \dfrac{(23.14 \text{ mL soln})(0.1200 \text{ N})}{(21.20 \text{ mL soln})} = 0.1310 \text{ N}$

15.23 $N_a = \dfrac{V_b \cdot N_b}{V_a} = \dfrac{(40.00 \text{ mL soln})(0.1000 \text{ N})}{(52.00 \text{ mL soln})} = 0.07692$

15.24 $N_b = \dfrac{V_a \cdot N_a}{V_b} = \dfrac{(50.9 \text{ mL soln})(0.0400 \text{ N})}{(42.4 \text{ mL soln})} = 0.0480 \text{ N}$

15.25 (a) $[H^+] = 0.21 \text{ M}$ (b) $[OH^-] = 0.16 \text{ M}$ (c) $[H^+] = 3.7 \times 10^{-3} \text{ M}$
(d) $[OH^-] = 5.2 \times 10^{-4} \text{ M}$

15.26 (a) $[OH^-] = \dfrac{1.0 \times 10^{-14}}{1.0 \times 10^{-2}} = 1.0 \times 10^{-12}$ (b) $[H^+] = \dfrac{1.0 \times 10^{-14}}{1.0 \times 10^{-11}} = 1.0 \times 10^{-3}$

(c) $[OH^-] = \dfrac{1.0 \times 10^{-14}}{3.5 \times 10^{-8}} = 2.9 \times 10^{-7}$ (d) $[H^+] = \dfrac{1.0 \times 10^{-14}}{8.5 \times 10^{-1}} = 1.2 \times 10^{-14}$

15.27 (a) strongly basic (b) weakly acidic (c) strongly acidic (d) weakly basic

15.28 (a) 9.50   (b) 7.00   (c) 4.90   (d) 14.00

15.29 (a) $[H^+] = 1.0 \times 10^{-3}$ M   pH = 3.00   pOH = 11.00
(b) $[OH^-] = 1.0 \times 10^{-5}$ M   pOH = 5.00 pH = 9.00
(c) $[H^+] = 1.0 \times 10^{-4}$ M   pH = 4.00   pOH = 10.00
(d) $[OH^-] = 1.0 \times 10^{0}$ M pOH = 0.00 pH = 14.00

15.30 (a) $0.63 \text{ g HNO}_3 \left(\dfrac{1 \text{ mol HNO}_3}{63.0 \text{ g HNO}_3} \right) = 0.010 \text{ mol HNO}_3$

molarity = $\dfrac{0.010 \text{ mol HNO}_3}{1.0 \text{ L soln}}$ = 0.010 M HNO$_3$ $(1.0 \times 10^{-2}$ M HNO$_3)$
$[H^+] = 1.0 \times 10^{-2}$ M
pH = $- \log(1.0 \times 10^{-2})$ = $-$ ($-$ 2.00) = 2.00

(b) $2.8 \text{ g KOH} \left(\dfrac{1 \text{ mol KOH}}{56.1 \text{ g KOH}} \right) = 0.050 \text{ mol KOH}$

molarity = $\dfrac{0.050 \text{ mol KOH}}{0.500 \text{ L soln}}$ = 0.10 M KOH $(1.0 \times 10^{-1}$ M KOH)
$[OH^-] = 1.0 \times 10^{-1}$ M
pOH = $- \log(1.0 \times 10^{-1})$ = $-$ ($-$ 1.00) = 1.00
pH = 13.00

(c) $0.73 \text{ g HCl} \left(\dfrac{1 \text{ mol HCl}}{36.5 \text{ g HCl}} \right) = 0.020 \text{ mol HCl}$

molarity = $\dfrac{0.020 \text{ mol HCl}}{20.0 \text{ L soln}}$ = 1.0×10^{-3} M HCl
$[H^+] = 1.0 \times 10^{-3}$ M
pH = $- \log(1.0 \times 10^{-3})$ = $-$ ($-$ 3.00) = 3.00

(d) $0.10 \text{ g NaOH} \left(\dfrac{1 \text{ mol NaOH}}{40.0 \text{ g NaOH}} \right) = 0.0025 \text{ mol NaOH}$

molarity = $\dfrac{0.0025 \text{ mol NaOH}}{25 \text{ L soln}}$ = 1.0×10^{-4} M NaOH
$[OH^-] = 1.0 \times 10^{-4}$ M
pOH = $- \log(1.0 \times 10^{-4})$ = $-$ ($-$ 4.00) = 4.00
pH = 10.00

15.31 (a) $[H^+] = 1.6 \times 10^{-2}$ M
pH = $- \log(1.6 \times 10^{-2})$ = $-$ ($-$ 1.80)
pH = 1.80

15.31 (continued)

   (b) $[OH^-] = 3.7 \times 10^{-4}$ M
       pOH $= -\log(3.7 \times 10^{-4}) = -(-3.43)$
       pOH $= 3.43$ pH $= 10.57$

   (c) $[H^+] = 2.4 \times 10^{-1}$ M
       pH $= -\log(2.4 \times 10^{-1}) = -(-0.62)$
       pH $= 0.62$

   (d) $[OH^-] = 6.6 \times 10^{-6}$ M
       pOH $= -\log(6.6 \times 10^{-6}) = -(-5.18)$
       pOH $= 5.18$ pH $= 8.82$

15.32 (a) $2.5 \text{ g HBr} \left(\dfrac{1 \text{ mol HBr}}{80.9 \text{ g HBr}} \right) = 0.031 \text{ mol HBr}$

       $\text{molarity} = \dfrac{0.031 \text{ mol HBr}}{50.0 \text{ L soln}} = 6.2 \times 10^{-4} \text{ M HBr}$

       $[H^+] = 6.2 \times 10^{-4}$ M
       pH $= -\log(6.2 \times 10^{-4}) = -(-3.21) = 3.21$

   (b) $0.076 \text{ g NaOH} \left(\dfrac{1 \text{ mol NaOH}}{40.0 \text{ g NaOH}} \right) = 0.0019 \text{ mol NaOH}$

       $\text{molarity} = \dfrac{0.0019 \text{ mol NaOH}}{25 \text{ L soln}} = 7.6 \times 10^{-5} \text{ M NaOH}$

       $[OH^-] = 7.6 \times 10^{-5}$ M
       pOH $= -\log(7.6 \times 10^{-5}) = -(-4.12) = 4.12$
       pH $= 9.88$

   (c) $0.22 \text{ g HNO}_3 \left(\dfrac{1 \text{ mol HNO}_3}{63.0 \text{ g HNO}_3} \right) = 0.0035 \text{ mol HNO}_3$

       $\text{molarity} = \dfrac{0.0035 \text{ mol HNO}_3}{0.75 \text{ L soln}} = 4.7 \times 10^{-3} \text{ M HNO}_3$

       $[H^+] = 4.7 \times 10^{-3}$ M
       pH $= -\log(4.7 \times 10^{-3}) = -(-2.33) = 2.33$

   (d) $4.2 \times 10^{-3} \text{ g KOH} \left(\dfrac{1 \text{ mol KOH}}{56.1 \text{ g KOH}} \right) = 7.5 \times 10^{-5} \text{ mol KOH}$

       $\text{molarity} = \dfrac{7.5 \times 10^{-5} \text{ mol KOH}}{5.0 \text{ L soln}} = 1.5 \times 10^{-5} \text{ M KOH}$

       $[OH^-] = 1.5 \times 10^{-5}$ M
       pOH $= -\log(1.5 \times 10^{-5}) = -(-4.82) = 4.82$
       pH $= 9.18$

15.33

	pH	pOH	[H⁺]	[OH⁻]
(a)	3.27	10.73	5.4×10^{-4} M	1.9×10^{-11} M
(b)	8.40	5.60	4.0×10^{-9} M	2.5×10^{-6} M
(c)	5.24	8.76	5.8×10^{-6} M	1.7×10^{-9} M
(d)	9.86	4.14	1.4×10^{-10} M	7.2×10^{-5} M

15.34 (a) 3 (b) 4 (c) 6 (d) 5 (e) 1 (f) 2

15.35 (a) $CN^- + H_2O \rightleftharpoons HCN + OH^-$

(b) $F^- + H_2O \rightleftharpoons HF + OH^-$

(c) $C_2H_3O_2^- + H_2O \rightleftharpoons HC_2H_3O_2 + OH^-$

(d) $CO_3^{2-} + H_2O \rightleftharpoons HCO_3^- + OH^-$

15.36

	acid	base
(a)	HBr	$C_2H_3O_2^-$
(b)	H_2O	NO_2^-
(c)	$HClO_4$	H_2O
(d)	HSO_4^-	OH^-

15.37 (a) NO_3^- (b) NH_4^+ (c) HSO_4^-

(d) CO_3^{2-} (e) $H_2PO_4^-$ (f) H_2CO_3 or CO_2(aq)

15.38 (a) $H_2C_2O_4$ (b) $C_2O_4^{2-}$

15.39

	acid	base
(a)	$AlCl_3$	NH_3
(b)	Ag^+	NH_3
(c)	Cu^+	CN^-
(d)	Pd^{2+}	Cl^-

15.40 (a) A weak acid is one that only partially dissociates.

(b) Whereas a weak acid is one that only partially dissociates, a strong acid is one that completely dissociates.

(c) A solution that is weakly acidic is one with a pH in the range of 5 to 7. Such a solution could be prepared by making a very dilute solution of a strong acid. A solution of a weak acid is any solution prepared from a weak acid.

15.41 [Remember: The solution that is the most acidic is the one with the lowest pH.]

(a) 0.10 M HF < 0.10 M $HC_7H_5O_2$ < 0.10 M $HC_2H_3O_2$

(b) 0.10 M HCl < 0.10 M HClO < 0.10 M HCN

(c) 0.10 M HNO_3 < 0.10 M HNO_2 < 0.010 M HNO_2

15.42 (a) no (b) yes (c) yes

15.43 The components of (b) and (c) are derived from strong acids, and therefore cannot be used to prepare buffers.

15.44 (a) 1 (b) 2 (c) 1 (d) 2 (e) 4

15.45 (a) $0.255 \text{ mol HClO}_4 \left(\dfrac{1 \text{ equiv HClO}_4}{1 \text{ mol HClO}_4} \right) = 0.255 \text{ equiv HClO}_4$

(b) $0.450 \text{ mol H}_3\text{PO}_4 \left(\dfrac{3 \text{ equiv H}_3\text{PO}_4}{1 \text{ mol H}_3\text{PO}_4} \right) = 1.35 \text{ equiv H}_3\text{PO}_4$

(c) $0.125 \text{ mol Ca(OH)}_2 \left(\dfrac{2 \text{ equiv Ca(OH)}_2}{1 \text{ mol Ca(OH)}_2} \right) = 0.250 \text{ equiv Ca(OH)}_2$

(d) $0.735 \text{ mol H}_2\text{C}_2\text{O}_4 \left(\dfrac{2 \text{ equiv H}_2\text{C}_2\text{O}_4}{1 \text{ mol H}_2\text{C}_2\text{O}_4} \right) = 1.47 \text{ equiv H}_2\text{C}_2\text{O}_4$

15.46 (a) $0.275 \text{ equiv HI} \left(\dfrac{1 \text{ mol HI}}{1 \text{ equiv HI}} \right) = 0.275 \text{ mol HI}$

(b) $0.834 \text{ equiv H}_3\text{PO}_4 \left(\dfrac{1 \text{ mol H}_3\text{PO}_4}{3 \text{ equiv H}_3\text{PO}_4} \right) = 0.278 \text{ mol H}_3\text{PO}_4$

(c) $0.510 \text{ equiv Al(OH)}_3 \left(\dfrac{1 \text{ mol Al(OH)}_3}{3 \text{ equiv Al(OH)}_3} \right) = 0.170 \text{ mol Al(OH)}_3$

(d) $2.44 \text{ equiv H}_2\text{SO}_4 \left(\dfrac{1 \text{ mol H}_2\text{SO}_4}{2 \text{ equiv H}_2\text{SO}_4} \right) = 1.22 \text{ mol H}_2\text{SO}_4$

15.47 (a) $4.32 \text{ g HNO}_3 \left(\dfrac{1 \text{ mol HNO}_3}{63.0 \text{ g HNO}_3} \right)\left(\dfrac{1 \text{ equiv HNO}_3}{1 \text{ mol HNO}_3} \right) = 0.0686 \text{ equiv HNO}_3$

(b) $1.00 \text{ g H}_2\text{C}_2\text{O}_4 \left(\dfrac{1 \text{ mol H}_2\text{C}_2\text{O}_4}{90.0 \text{ g H}_2\text{C}_2\text{O}_4} \right)\left(\dfrac{2 \text{ equiv H}_2\text{C}_2\text{O}_4}{1 \text{ mol H}_2\text{C}_2\text{O}_4} \right) = 0.0222 \text{ equiv H}_2\text{C}_2\text{O}_4$

(c) $0.935 \text{ g H}_3\text{PO}_4 \left(\dfrac{1 \text{ mol H}_3\text{PO}_4}{98.0 \text{ g H}_3\text{PO}_4} \right)\left(\dfrac{3 \text{ equiv H}_3\text{PO}_4}{1 \text{ mol H}_3\text{PO}_4} \right) = 0.0286 \text{ equiv H}_3\text{PO}_4$

(d) $0.0352 \text{ g Mg(OH)}_2 \left(\dfrac{1 \text{ mol Mg(OH)}_2}{58.3 \text{ g Mg(OH)}_2} \right)\left(\dfrac{2 \text{ equiv Mg(OH)}_2}{1 \text{ mol Mg(OH)}_2} \right)$
$= 0.00121 \text{ equiv Mg(OH)}_2$

15.48 (a) $5.00 \text{ g KOH} \left(\dfrac{1 \text{ mol KOH}}{56.1 \text{ g KOH}} \right) \left(\dfrac{1 \text{ equiv KOH}}{1 \text{ mol KOH}} \right) = 0.0891 \text{ equiv KOH}$

$\qquad \text{normality} = \dfrac{0.0891 \text{ equiv KOH}}{0.150 \text{ L soln}} = 0.594 \text{ N KOH}$

(b) $0.525 \text{ g H}_2\text{SO}_4 \left(\dfrac{1 \text{ mol H}_2\text{SO}_4}{98.1 \text{ g H}_2\text{SO}_4} \right) \left(\dfrac{2 \text{ equiv H}_2\text{SO}_4}{1 \text{ mol H}_2\text{SO}_4} \right) = 0.0107 \text{ equiv H}_2\text{SO}_4$

$\qquad \text{normality} = \dfrac{0.0107 \text{ equiv H}_2\text{SO}_4}{0.0750 \text{ L soln}} = 0.143 \text{ N H}_2\text{SO}_4$

(c) $0.115 \text{ g Ca(OH)}_2 \left(\dfrac{1 \text{ mol Ca(OH)}_2}{74.1 \text{ g Ca(OH)}_2} \right) \left(\dfrac{2 \text{ equiv Ca(OH)}_2}{1 \text{ mol Ca(OH)}_2} \right) = 0.00310 \text{ equiv Ca(OH)}_2$

$\qquad \text{normality} = \dfrac{0.00310 \text{ equiv Ca(OH)}_2}{0.250 \text{ L soln}} = 0.0124 \text{ N Ca(OH)}_2$

(d) $7.55 \text{ g H}_3\text{PO}_4 \left(\dfrac{1 \text{ mol H}_3\text{PO}_4}{98.0 \text{ g H}_3\text{PO}_4} \right) \left(\dfrac{3 \text{ equiv H}_3\text{PO}_4}{1 \text{ mol H}_3\text{PO}_4} \right) = 0.231 \text{ equiv H}_3\text{PO}_4$

$\qquad \text{normality} = \dfrac{0.231 \text{ equiv H}_3\text{PO}_4}{1.50 \text{ L soln}} = 0.154 \text{ N H}_3\text{PO}_4$

15.49 (a) $0.250 \text{ L soln} \left(\dfrac{1.24 \text{ equiv H}_3\text{PO}_4}{\text{L soln}} \right) = 0.310 \text{ equiv H}_3\text{PO}_4$

(b) $0.0400 \text{ L soln} \left(\dfrac{6.00 \text{ equiv NaOH}}{\text{L soln}} \right) = 0.240 \text{ equiv NaOH}$

(c) $0.375 \text{ L soln} \left(\dfrac{0.350 \text{ equiv HCl}}{\text{L soln}} \right) = 0.131 \text{ equiv HCl}$

(d) $0.0050 \text{ L soln} \left(\dfrac{12 \text{ equiv H}_3\text{PO}_4}{\text{L soln}} \right) = 0.060 \text{ equiv H}_3\text{PO}_4$

(e) $0.0750 \text{ L soln} \left(\dfrac{0.500 \text{ equiv H}_2\text{C}_2\text{O}_4}{\text{L soln}} \right) = 0.0375 \text{ equiv H}_2\text{C}_2\text{O}_4$

15.50 $N_b = \dfrac{V_a \cdot N_a}{V_b} = \dfrac{(20.0 \text{ mL soln})(0.215 \text{ N})}{(35.0 \text{ mL soln})} = 0.123 \text{ N}$

15.51 $N_a = \dfrac{V_b \cdot N_b}{V_a} = \dfrac{(36.5 \text{ mL soln})(0.355 \text{ N})}{(22.0 \text{ mL soln})} = 0.589 \text{ N}$

15.52 $N_a = \dfrac{V_b \cdot N_b}{V_a} = \dfrac{(12.5 \text{ mL soln})(0.344 \text{ N})}{(17.4 \text{ mL soln})} = 0.247 \text{ N}$

15.53 In a neutral solution $[H^+] = [OH^-]$.

Since $[H^+][OH^-] = K_w$, in a neutral solution $[H^+] = \sqrt{K_w}$.

If K_w were 1.0×10^{-16}, $[H^+]$ would be 1.0×10^{-8}, and the pH $= -\log(1.0 \times 10^{-8}) = 8.00$.

15.54 The fraction that corresponds to 0.0063% is 0.000063 (or 6.3×10^{-5}). Since that is the fraction of HCN molecules that dissociate, the hydrogen ion concentration is:

$(6.3 \times 10^{-5})(0.10 \text{ M}) = 6.3 \times 10^{-6} \text{ M H}^+$

pH $= -\log(6.3 \times 10^{-6}) = -(-5.20) = 5.20$

15.55 Blood that is entering the lungs carries carbon dioxide to be expelled from the body. Since carbon dioxide behaves as an acid when dissolved in water, it lowers the pH. Thus, the blood pH is lower before the blood enters the lungs and higher after it leaves.

15.56 Sodium hydroxide absorbs carbon dioxide from the atmospHere. When dissolved in water, carbon dioxide neutralizes some of the hydroxide ions as follows:

$CO_2 + OH^- \longrightarrow HCO_3^-$.

15.57 (a) pH = 5   (b) pH = = 4   (c) pH = 6
(d) Yes. It must have a pH less than 4.
(e) No. It could have a pH as low as 5, which is in the acid region.
(f) No. Bromcresol green will not turn yellow above a pH of 4, and methyl red will not turn yellow below a pH of 6. Thus, no single solution could turn both of them yellow.

15.58 (a) normality $= \dfrac{0.240 \text{ mol H}_2\text{SO}_4}{\text{L soln}}\left(\dfrac{2 \text{ equiv H}_2\text{SO}_4}{1 \text{ mol H}_2\text{SO}_4}\right) = \dfrac{0.480 \text{ equiv H}_2\text{SO}_4}{\text{L soln}}$

$= 0.480 \text{ N H}_2\text{SO}_4$

(b) normality $= \dfrac{0.015 \text{ mol Ca(OH)}_2}{\text{L soln}}\left(\dfrac{2 \text{ equiv Ca(OH)}_2}{1 \text{ mol Ca(OH)}_2}\right) = \dfrac{0.030 \text{ equiv Ca(OH)}_2}{\text{L soln}}$

$= 0.030 \text{ N Ca(OH)}_2$

(c) normality $= \dfrac{0.350 \text{ equiv HNO}_3}{\text{L soln}}\left(\dfrac{1 \text{ mol HNO}_3}{1 \text{ equiv HNO}_3}\right) = \dfrac{0.350 \text{ mol HNO}_3}{\text{L soln}}$

$= 0.350 \text{ M HNO}_3$

(d) normality $= \dfrac{0.420 \text{ equiv H}_3\text{PO}_4}{\text{L soln}}\left(\dfrac{1 \text{ mol H}_3\text{PO}_4}{3 \text{ equiv H}_3\text{PO}_4}\right) = \dfrac{0.140 \text{ mol H}_3\text{PO}_4}{\text{L soln}}$

$= 0.140 \text{ M H}_3\text{PO}_4$

15.59 (a) $[Cu^{2+}] = 0.25$ M; $[NO_3^-] = 0.50$ M

(b) $[Mg^{2+}] = 0.35$ M; $[Br^-] = 0.70$ M

(c) $[K^+] = 0.36$ M; $[PO_4^{3-}] = 0.12$ M

(d) $[Al^{3+}] = 0.44$ M; $[SO_4^{2-}] = 0.66$ M

15.60 (a) 2 equiv = 1 mol

(b) $5.78 \text{ g H}_3\text{PO}_4 \left(\dfrac{1 \text{ mol H}_3\text{PO}_4}{98.0 \text{ g H}_3\text{PO}_4} \right) \left(\dfrac{2 \text{ equiv H}_3\text{PO}_4}{1 \text{ mol H}_3\text{PO}_4} \right) = 0.118 \text{ equiv H}_3\text{PO}_4$

$\text{normality} = \dfrac{0.118 \text{ equiv H}_3\text{PO}_4}{0.500 \text{ L soln}} = 0.236 \text{ N H}_3\text{PO}_4$

(c) $V_a \cdot N_a = V_b \cdot N_b$

$(25.4 \text{ mL})(0.236 \text{ N}) = (16.8 \text{ mL})(N_b)$

$\dfrac{(25.4 \text{ mL})(0.236 \text{ N})}{(16.8 \text{ mL})} = N_b = 0.357 \text{ N}$

CHAPTER 16

Oxidation-Reduction Reactions

16.1 (a) oxidation (b) reduction (c) oxidation (d) reduction

16.2 (a) a loss of electrons   (b) a gain of electrons
(c) either the oxidation or the reduction portion of an oxidation–reduction
(d) a substance that causes oxidation   (e) a substance that causes reduction

16.3 Key: o.a. = oxidizing agent;   r.a. = reducing agent
(a) o.a. = Cu^{2+}; r.a. = Mg   (b) o.a. = Br_2; r.a. = I^-
(c) o.a. = Cl_2;   r.a. = Na

16.4 (a) H = 0   (b) Al = +3
(c) H = + 1; P = + 5; O = − 2   (d) N = + 2; O = − 2
(e) N = + 5; O = − 2   (f) H = + 1; N = + 3; O = − 2
(g) Cr = + 6; O = − 2   (h) Mn = + 4; O = − 2
(i) Mn = + 7; O = − 2   (j) C = + 3; O = − 2

16.5 (a) F = 0   (b) C = + 4; O = − 2
(c) Br = + 3; O = − 2   (d) K = + 1; I = − 1
(e) Cu = + 2; Br = − 1   (f) Li = + 1; N = + 5; O = − 2
(g) Na = + 1; S = + 6; O = − 2   (h) Al = + 3; S = + 6; O = − 2

16.6 (a) $\overset{0}{Zn} + \overset{+1 \; +6 \; -2}{H_2SO_4} \longrightarrow \overset{+2 \; +6 \; -2}{ZnSO_4} + \overset{0}{H_2}$
Yes. Zinc is oxidized; hydrogen is reduced.

(b) $\overset{+1 \; -1}{KI} + \overset{+2 \quad +5 \; -2}{Pb(NO_3)_2} \longrightarrow \overset{+2 \; -1}{PbI_2} + \overset{+1 \; +5 \; -2}{KNO_3}$
No changes.

(c) $\overset{0}{Mn} + \overset{+1 \; -1}{HCl} \longrightarrow \overset{+2 \quad -1}{MnCl_2} + \overset{0}{H_2}$
Yes. Manganese is oxidized; hydrogen is reduced.

(d) $\overset{+2 \; -2}{HgO} + \overset{+1 \; +5 \; -2}{HNO_3} \longrightarrow \overset{+2 \quad +5 \; -2}{Hg(NO_3)_2} + \overset{+1 \; -2}{H_2O}$
No changes.

16.7 $\overset{0}{Cu} + \overset{+1 \; +5 \; -2}{HNO_3} \longrightarrow \overset{+2 \quad +5 \; -2}{Cu(NO_3)_2} + \overset{+2 \; -2}{NO} + \overset{+1 \; -2}{H_2O}$
Copper is oxidized from 0 to +2. Nitrogen is reduced from +5 to +2. However, only a portion of the nitrogen is reduced.

16.8 (a) $\overset{0}{Sn} + 4 \overset{+5}{HNO_3} \longrightarrow \overset{+4}{SnO_2} + 4 \overset{+4}{NO_2} + 2 H_2O$
Each Sn atom loses 4 e^-; each N atom gains 1 e^-.

16.8 (continued)

(b) $\overset{-1}{2\,HBr} + \overset{+6}{H_2\,SO_4} \longrightarrow \overset{+4}{SO_2} + \overset{0}{Br_2} + 2\,H_2O$

Each Br atom loses 1 e^-; each S atom gains 2 e^-.

(c) $\overset{+5}{2\,HNO_3} + \overset{-1}{6\,HCl} \longrightarrow \overset{+2}{2\,NO} + \overset{0}{3\,Cl_2} + 4\,H_2O$

Each Cl atom loses 1 e^-; each N atom gains 3 e^-.

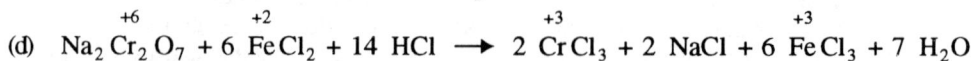

(d) $\overset{+6}{Na_2\,Cr_2\,O_7} + \overset{+2}{6\,FeCl_2} + 14\,HCl \longrightarrow \overset{+3}{2\,CrCl_3} + 2\,NaCl + \overset{+3}{6\,FeCl_3} + 7\,H_2O$

Each Fe atom loses 1 e^-; each Cr atom gains 3 e^-.

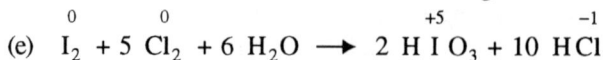

(e) $\overset{0}{I_2} + \overset{0}{5\,Cl_2} + 6\,H_2O \longrightarrow \overset{+5}{2\,HIO_3} + \overset{-1}{10\,HCl}$

Each I atom loses 5 e^-; each Cl atom gains 1 e^-.

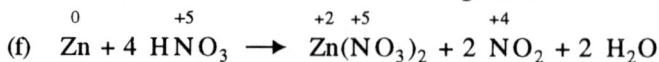

(f) $\overset{0}{Zn} + \overset{+5}{4\,HNO_3} \longrightarrow \overset{+2}{Zn(NO_3)_2} + \overset{+5}{2\,NO_2} + 2\,H_2O$

Each Zn atom loses 2 e^-; two N atoms gain 1 e^- each; two N atoms do not change.

16.9 (a) $Sn^{2+} \longrightarrow Sn^{4+} + 2\,e^-$ (oxidation)

(b) $MnO_4^- + 8\,H^+ + 5\,e^- \longrightarrow Mn^{2+} + 4\,H_2O$ (reduction)

(c) $C_2O_4^{2-} \longrightarrow 2\,CO_2 + 2\,e^-$ (oxidation)

(d) $2\,I^- \longrightarrow I_2 + 2\,e^-$ (oxidation)

(e) $Cr_2O_7^{2-} + 14\,H^+ + 6\,e^- \longrightarrow 2\,Cr^{3+} + 7\,H_2O$ (reduction)

16.10 (a)

$$3(Cu \longrightarrow Cu^{2+} + 2\,e^-)$$ (oxidation)

$$2(3\,e^- + 4\,H^+ + NO_3^- \longrightarrow NO + 2\,H_2O)$$ (reduction)

$$3\,Cu + 2\,NO_3^- + 8\,H^+ \longrightarrow 3\,Cu^{2+} + 2\,NO + 4\,H_2O$$

(b)

$$3(4\,H_2O + S^{2-} \longrightarrow SO_4^{2-} + 8\,H^+ + 8\,e^-)$$ (oxidation)

$$8(3\,e^- + 4\,H^+ + NO_3^- \longrightarrow NO + 2\,H_2O)$$ (reduction)

$$8\,NO_3^- + 3\,S^{2-} + 8\,H^+ \longrightarrow 8\,NO + 3\,SO_4^{2-} + 4\,H_2O$$

(c)

$$5(2\,H_2O + NO \longrightarrow NO_3^- + 4\,H^+ + 3\,e^-)$$ (oxidation)

$$3(5\,e^- + 8\,H^+ + MnO_4^- \longrightarrow Mn^{2+} + 4\,H_2O)$$ (reduction)

$$3\,MnO_4^- + 5\,NO + 4\,H^+ \longrightarrow 3\,Mn^{2+} + 5\,NO_3^- + 2\,H_2O$$

16.10 (continued)

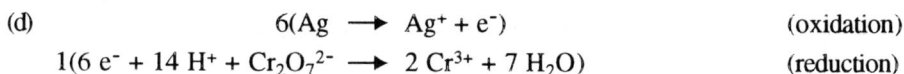

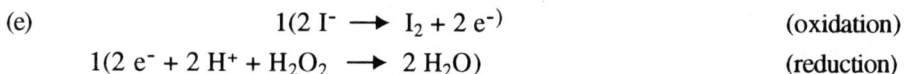

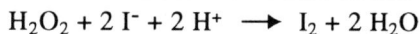

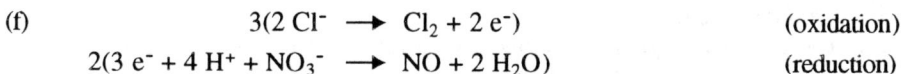

(d) $6(Ag \longrightarrow Ag^+ + e^-)$ (oxidation)

$\underline{1(6\ e^- + 14\ H^+ + Cr_2O_7{}^{2-} \longrightarrow 2\ Cr^{3+} + 7\ H_2O)}$ (reduction)

$6\ Ag + Cr_2O_7{}^{2-} + 14\ H^+ \longrightarrow 6\ Ag^+ + 2\ Cr^{3+} + 7\ H_2O$

(e) $1(2\ I^- \longrightarrow I_2 + 2\ e^-)$ (oxidation)

$\underline{1(2\ e^- + 2\ H^+ + H_2O_2 \longrightarrow 2\ H_2O)}$ (reduction)

$H_2O_2 + 2\ I^- + 2\ H^+ \longrightarrow I_2 + 2\ H_2O$

(f) $3(2\ Cl^- \longrightarrow Cl_2 + 2\ e^-)$ (oxidation)

$\underline{2(3\ e^- + 4\ H^+ + NO_3{}^- \longrightarrow NO + 2\ H_2O)}$ (reduction)

$2\ NO_3{}^- + 6\ Cl^- + 8\ H^+ \longrightarrow 2\ NO + 3\ Cl_2 + 4\ H_2O$

16.11 Key: soa = stronger oxidizing agent   sra = stronger reducing agent
                      woa = weaker oxidizing agent   wra = weaker reducing agent

(a) Ba + $Cd^{2+} \longrightarrow Ba^{2+}$ + Cd
     (sra)   (soa)   (woa)   (wra)   Yes. Reaction occurs.

(b) 3 Cu + $2\ Al^{3+} \longrightarrow 3\ Cu^{2+}$ + 2 Al
     (wra)   (woa)   (soa)   (sra)   No reaction.

(c) Mg + $2\ H^+ \longrightarrow Mg^{2+}$ + H_2
     (sra)   (soa)   (woa)   (wra)   Yes. Reaction occurs.

(d) 2 Au + $6\ H^+ \longrightarrow 2\ Au^{3+}$ + $3\ H_2$
     (wra)   (woa)   (soa)   (sra)   No reaction.

(e) Pb + $2\ H^+ \longrightarrow Pb^{2+}$ + H_2
     (sra)   (soa)   (woa)   (wra)   Yes. Reaction occurs.

16.12 (a) An electrochemical cell is an arrangement of two half-reactions such that the electrons transferred in the oxidation-reduction reaction must pass through an external circuit from one electrode to another.

(b) A voltaic cell is an electrochemical cell that operates in the spontaneous direction. An electrolytic cell is connected to an external source of energy and runs in the nonspontaneous direction.

(c) Oxidation occurs at the anode.

(d) Reduction occurs at the cathode.

(e) The porous partition prevents the solutions in the two half-cells from mixing, but it enables ions to migrate in order to maintain the electrical neutrality in each half-cell.

(f) Anions migrate toward the anode.

16.13 (a) $Cu \longrightarrow Cu^{2+} + 2\ e^-$ (b) The spoon is the anode.

(c) The spoon dissolves.

16.14 $2 PbSO_4(s) + 2 H_2O(l) \longrightarrow Pb(s) + PbO_2(s) + 2 H_2SO_4(aq)$
During recharging, the battery operates as an electrolytic cell.

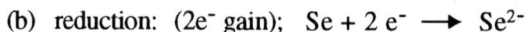

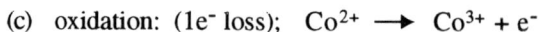

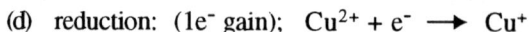

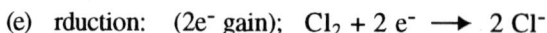

16.15 (a) oxidation: ($1e^-$ loss); $Na \longrightarrow Na^+ + e^-$
   (b) reduction: ($2e^-$ gain); $Se + 2 e^- \longrightarrow Se^{2-}$
   (c) oxidation: ($1e^-$ loss); $Co^{2+} \longrightarrow Co^{3+} + e^-$
   (d) reduction: ($1e^-$ gain); $Cu^{2+} + e^- \longrightarrow Cu^+$
   (e) rduction: ($2e^-$ gain); $Cl_2 + 2 e^- \longrightarrow 2 Cl^-$

16.16 (a) The electrons from an oxidation must go somewhere, and the species that acquires them must be reduced.
   (b) The oxidizing agent takes electrons away from the species it oxidizes. Since it is gaining electrons, it must be reduced.
   (c) The reducing agent must give electrons to the species it reduces. Since it is losing electrons, it must be oxidized.

16.17 Key: o.a. = oxidizing agent;   r.a. = reducing agent
   (a) o.a. = H^+ r.a. = Mg   (b) o.a. = Cl_2 r.a. = Br^-
   (c) o.a. = I_2 r.a. = K

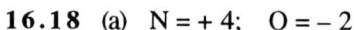

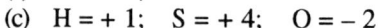

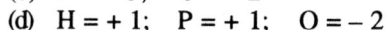

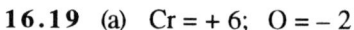

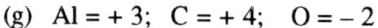

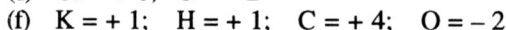

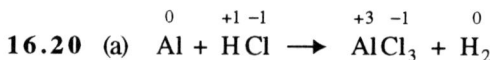

16.18 (a) N = + 4; O = − 2   (b) P = + 5;   O = − 2
   (c) H = + 1; S = + 4;   O = − 2   (d) H = + 1;   P = + 1;   O = − 2
   (e) Tl = + 1; O = − 2   (f) H = + 1;   Cl = + 1; O = − 2
   (g) Zr = + 3; Cl = − 1   (h) K = + 1;   Mn = + 7;   O = − 2

16.19 (a) Cr = + 6; O = − 2   (b) H = + 1;   I = + 7; O = − 2
   (c) K = + 1; Se = − 2   (d) Cr = + 6; O = − 2
   (e) U = + 6; F = − 1   (f) K = + 1;   H = + 1;   C = + 4;   O = − 2
   (g) Al = + 3; C = + 4;   O = − 2   (h) S = + 2;   O = − 2

16.20 (a) $\overset{0}{Al} + \overset{+1\ -1}{HCl} \longrightarrow \overset{+3\ -1}{AlCl_3} + \overset{0}{H_2}$
   Oxidation-reduction occurs: aluminum goes from zero to +3; hydrogen goes from +1 to zero.

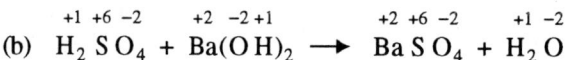

   (b) $\overset{+1\ +6\ -2}{H_2SO_4} + \overset{+2\ -2+1}{Ba(OH)_2} \longrightarrow \overset{+2\ +6\ -2}{BaSO_4} + \overset{+1\ -2}{H_2O}$
   This is not an oxidation-reduction.

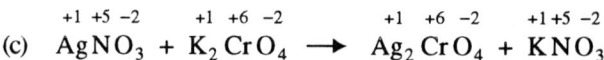

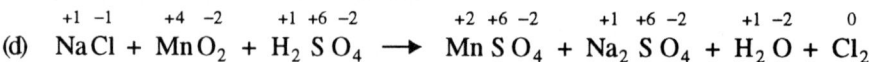

   (c) $\overset{+1\ +5\ -2}{AgNO_3} + \overset{+1\ +6\ -2}{K_2CrO_4} \longrightarrow \overset{+1\ +6\ -2}{Ag_2CrO_4} + \overset{+1+5\ -2}{KNO_3}$
   This is not an oxidation-reduction.

   (d) $\overset{+1\ -1}{NaCl} + \overset{+4\ -2}{MnO_2} + \overset{+1\ +6\ -2}{H_2SO_4} \longrightarrow \overset{+2\ +6\ -2}{MnSO_4} + \overset{+1\ +6\ -2}{Na_2SO_4} + \overset{+1\ -2}{H_2O} + \overset{0}{Cl_2}$
   Oxidation-reduction occurs: manganese goes from +4 to +2; chlorine goes from −1 to zero.

16.21 (a) $\overset{+5}{8\ HNO_3} + 3\ \overset{-2}{CuS} \longrightarrow 3\ \overset{+6}{CuSO_4} + 8\ \overset{+2}{NO} + 4\ H_2O$

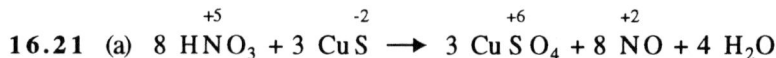

N gains 3 e^- per atom (from +5 to +2)

S loses 8 e^- per atom (from -2 to +6)

(b) $\overset{0}{6\ Sb} + \overset{+5}{10\ HNO_3} \longrightarrow 3\ \overset{+5}{Sb_2O_5} + 10\ \overset{+2}{NO} + 5\ H_2O$

N gains 3 e^- per atom (from +5 to +2)

Sb loses 5 e^- per atom (from zero to +5)

(c) $\overset{+3}{AuCl_3} + 2\ \overset{-1}{KI} \longrightarrow \overset{+1}{AuCl} + \overset{0}{I_2} + 2\ KCl$

Au gains 2 e^- per atom (from +3 to +1)

I loses 1 e^- per atom (from -1 to zero)

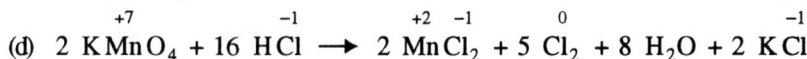

(d) $2\ \overset{+7}{KMnO_4} + 16\ \overset{-1}{HCl} \longrightarrow 2\ \overset{+2\ -1}{MnCl_2} + 5\ \overset{0}{Cl_2} + 8\ H_2O + 2\ \overset{-1}{KCl}$

Mn gains 5 e^- per atom (from +7 to +2)

10 Cl lose 1 e^- per atom (from -1 to zero); 6 Cl^- are spectator ions

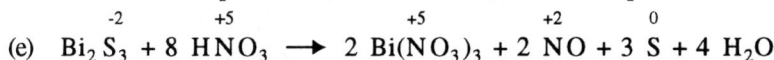

(e) $\overset{-2}{Bi_2S_3} + 8\ \overset{+5}{HNO_3} \longrightarrow 2\ \overset{+5}{Bi(NO_3)_3} + 2\ \overset{+2}{NO} + 3\ \overset{0}{S} + 4\ H_2O$

2 N gain 3 e^- per atom (from +5 to +2); 6 NO_3^- are spectator ions

S loses 2 e^- per atom (from -2 to zero)

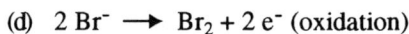

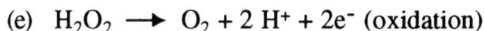

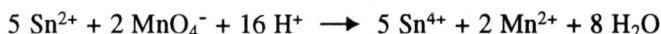

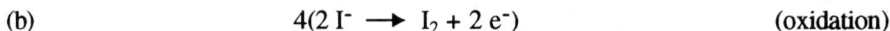

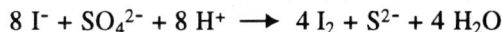

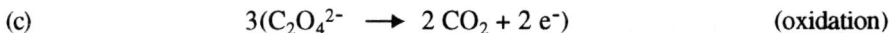

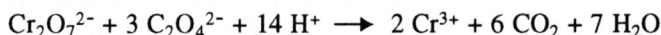

16.22 (a) $Cu^+ \longrightarrow Cu^{2+} + e^-$ (oxidation)

(b) $NO_3^- + 4\ H^+ + 3\ e^- \longrightarrow NO + 2\ H_2O$ (reduction)

(c) $2\ S_2O_3^{2-} \longrightarrow S_4O_6^{2-} + 2\ e^-$ (oxidation)

(d) $2\ Br^- \longrightarrow Br_2 + 2\ e^-$ (oxidation)

(e) $H_2O_2 \longrightarrow O_2 + 2\ H^+ + 2e^-$ (oxidation)

16.23 (a)

$5\ (Sn^{2+} \longrightarrow Sn^{4+} + 2\ e^-)$ (oxidation)

$2\ (MnO_4^- + 8\ H^+ + 5\ e^- \longrightarrow Mn^{2+} + 4\ H_2O)$ (reduction)

$\overline{}$

$5\ Sn^{2+} + 2\ MnO_4^- + 16\ H^+ \longrightarrow 5\ Sn^{4+} + 2\ Mn^{2+} + 8\ H_2O$

(b)

$4(2\ I^- \longrightarrow I_2 + 2\ e^-)$ (oxidation)

$1(SO_4^{2-} + 8\ H^+ + 8\ e^- \longrightarrow S^{2-} + 4\ H_2O)$ (reduction)

$\overline{}$

$8\ I^- + SO_4^{2-} + 8\ H^+ \longrightarrow 4\ I_2 + S^{2-} + 4\ H_2O$

(c)

$3(C_2O_4^{2-} \longrightarrow 2\ CO_2 + 2\ e^-)$ (oxidation)

$1(Cr_2O_7^{2-} + 14\ H^+ + 6\ e^- \longrightarrow 2\ Cr^{3+} + 7\ H_2O)$ (reduction)

$\overline{}$

$Cr_2O_7^{2-} + 3\ C_2O_4^{2-} + 14\ H^+ \longrightarrow 2\ Cr^{3+} + 6\ CO_2 + 7\ H_2O$

16.23 (continued)

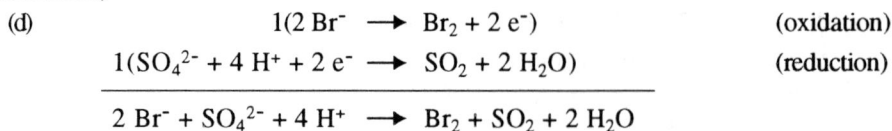

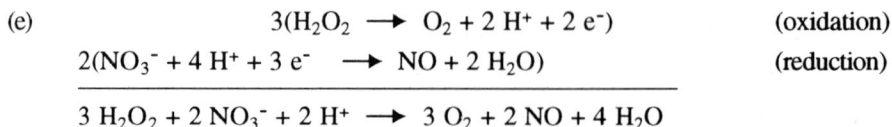

(d)
$$1(2\,Br^- \longrightarrow Br_2 + 2\,e^-) \quad \text{(oxidation)}$$
$$1(SO_4^{2-} + 4\,H^+ + 2\,e^- \longrightarrow SO_2 + 2\,H_2O) \quad \text{(reduction)}$$
$$2\,Br^- + SO_4^{2-} + 4\,H^+ \longrightarrow Br_2 + SO_2 + 2\,H_2O$$

(e)
$$3(H_2O_2 \longrightarrow O_2 + 2\,H^+ + 2\,e^-) \quad \text{(oxidation)}$$
$$2(NO_3^- + 4\,H^+ + 3\,e^- \longrightarrow NO + 2\,H_2O) \quad \text{(reduction)}$$
$$3\,H_2O_2 + 2\,NO_3^- + 2\,H^+ \longrightarrow 3\,O_2 + 2\,NO + 4\,H_2O$$

16.24

	reducing agent		oxidizing agent				
(a)	Ba	+	Fe^{2+}	$\longrightarrow$	Ba^{2+}	+ Fe	(spontaneous)
(b)	Cu	+	Cd^{2+}	$\longrightarrow$	Cu^{2+}	+ Cd	(nonspontaneous)
(c)	3 Mg	+	$2\,Al^{3+}$	$\longrightarrow$	$3\,Mg^{2+}$	+ 2 Al	(spontaneous)
(d)	2 Au	+	$3\,Cu^{2+}$	$\longrightarrow$	$2\,Au^{3+}$	+ 3 Cu	(nonspontaneous)
(e)	2 K	+	Zn^{2+}	$\longrightarrow$	$2\,K^+$	+ Zn	(spontaneous)
(f)	H_2	+	Hg_2^{2+}	$\longrightarrow$	$2\,H^+$	+ 2 Hg	(spontaneous)

16.25 (a) Pb^{2+} (b) $Mg + Pb^{2+} \longrightarrow Mg^{2+} + Pb$
(c) To the right.   (d) To the left.
(e) Mg   (f) Pb
(g) Mg is the anode; Pb is the cathode.

16.26 (a) right; $Na^+ + e^- \longrightarrow Na$; cathode.   (b) left; $2\,Cl^- \longrightarrow Cl_2 + 2\,e^-$; anode.

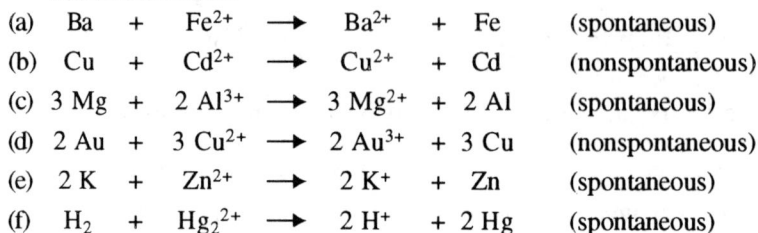

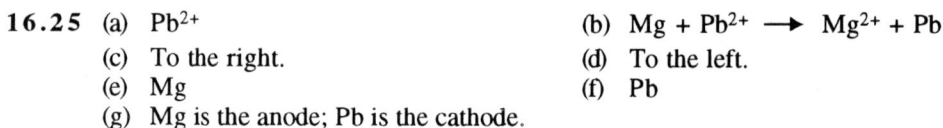

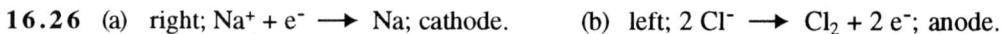

16.27 (a) As the battery operates, electrons from the lead electrode travel through the external circuit to the lead(IV) oxide electrode. Lead(II) ions are formed at each electrode, where they combine with sulfate ions from the sulfuric acid solution present.
(b) The "water" is really sulfuric acid, which is needed as a reactant.
(c) When the battery is recharged, this voltaic cell is converted into an electrolytic cell, thereby reversing the reaction and regenerating the reactants. Recharging will not work indefinitely because the lead(II) sulfate formed at each electrode during its spontaneous operation may come loose from the electrode and be unavailable for regeneration to lead and lead(IV) oxide during recharging. Eventually the electrodes become too corroded for adequate functioning of the battery.

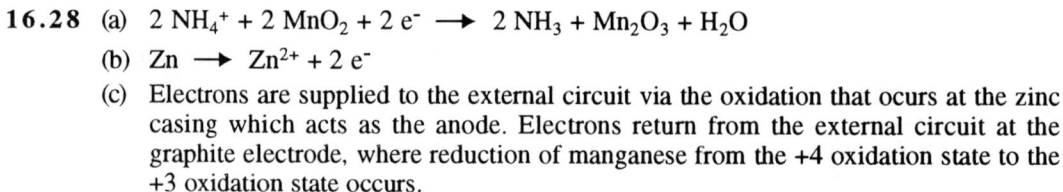

16.28 (a) $2\,NH_4^+ + 2\,MnO_2 + 2\,e^- \longrightarrow 2\,NH_3 + Mn_2O_3 + H_2O$
(b) $Zn \longrightarrow Zn^{2+} + 2\,e^-$
(c) Electrons are supplied to the external circuit via the oxidation that ocurs at the zinc casing which acts as the anode. Electrons return from the external circuit at the graphite electrode, where reduction of manganese from the +4 oxidation state to the +3 oxidation state occurs.

16.28 (continued)

(d) The graphite rod serves as an electrode.

(e) As ammonium ions (NH_4^+) are consumed to form uncharged ammonia molecules (NH_3), zinc ions (Zn^{2+}) are formed from zinc atoms (Zn), thereby replacing the positive charge.

(f) The chloride ions are spectator ions, needed to balance the positive charges of the ammonium ions and zinc ions.

(g) A zinc-carbon battery is not rechargeable.

16.29 (a) +6 in sulfate; +4 in sulfite.   (b) +5 in nitrate; +3 in nitrite.

(c) +5 in chlorate; +3 in chlorite.   (d) The oxidation state is higher in the *-ate* ion.

(e) It should have a higher oxidation state in phosphoric acid, since *-ic acids* come from *-ate* ions.

16.30 (a) $10\ HNO_3 + I_2 \longrightarrow 10\ NO_2 + 2\ HIO_3 + 4\ H_2O$

(b) $Cr_2O_7^{2-} + 3\ SO_2 + 2\ H^+ \longrightarrow 2\ Cr^{3+} + 3\ SO_4^{2-} + H_2O$

(c) $MnO_2 + 4\ HCl \longrightarrow MnCl_2 + 2\ Cl_2 + 2\ H_2O$

(d) $5\ C_2O_4^{2-} + 2\ MnO_4^- + 16\ H^+ \longrightarrow 10\ CO_2 + 2\ Mn^{2+} + 8\ H_2O$

(e) $3\ CuO + 2\ NH_3 \longrightarrow 3\ Cu + N_2 + 3\ H_2O$

(f) $IO_4^- + 7\ I^- + 8\ H^+ \longrightarrow 4\ I_2 + 4\ H_2O$

(g) $3\ KIO \longrightarrow KIO_3 + 2\ KI$

(h) $C_3H_8 + 5\ O_2 \longrightarrow 3\ CO_2 + 4\ H_2O$

[Note: Although part (h) is an oxidation-reduction, the method you learned in Chapter 10 is more appropriate for balancing this equation.]

16.31 (a) No. Hydrogen ions would oxidize the lead, thereby dissolving the vessel: $Pb + 2\ H^+ \longrightarrow Pb^{2+} + H_2$.

(b) Yes. Hydrogen ions will not oxidize copper: $Cu + H^+ \longrightarrow$ no reaction.

(c) Yes. Since hydrogen (H_2) is a stronger reducing agent than copper metal, the following reaction occurs: $Cu^{2+} + H_2 \longrightarrow Cu + 2\ H^+$.

(d) No. Magnesium is a stronger reducing agent than hydrogen. Thus, hydrogen will not reduce magnesium ions: $Mg^{2+} + H_2 \longrightarrow$ no reaction.

16.32 $I_2\ +\ 2\ e^- \rightleftharpoons 2\ I^-$ (I^- is the strongest reducing agent.)

$Br_2\ +\ 2\ e^- \rightleftharpoons 2\ Br^-$

$Cl_2\ +\ 2\ e^- \rightleftharpoons 2\ Cl^-$

$F_2\ +\ 2\ e^- \rightleftharpoons 2\ F^-$ (F_2 is the strongest oxidizing agent.)

16.33 $Al + 3\ Ag^+ \longrightarrow Al^{3+} + 3\ Ag$ The aluminum acts as the anode.

16.34 (a) $Cl_2 + H_2O \longrightarrow HClO + HCl$

(b) $Cl_2 + 2 H_2O \longrightarrow 2 HClO + 2 H^+ + 2 e^-$ (Cl_2 is the reducing agent.)

(c) $Cl_2 + 2 H^+ + 2 e^- \longrightarrow 2 HCl$ (Cl_2 is the oxidizing agent.)

(d) Addition of the equations in b and c gives:

$2 Cl_2 + 2 H_2O + 2 H^+ + 2 e^- \longrightarrow 2 HClO + 2 HCl + 2 H^+ + 2 e^-$

Elimination of 2 H^+ and 2 e^- from both sides gives:

$2 Cl_2 + 2 H_2O \longrightarrow 2 HClO + 2 HCl$

Division of all coeffiients by 2 gives the equation in a.

(e) Chlorine acts as both the oxidizing and reducing agent. This type of reaction is known as an autoxidation, since the same substance is both oxidized and reduced.)

16.35 Magnesium is a stronger reducing agent than iron. That means that it is more readily oxidized than iron. Because of this characteristic, if a magnesium rod is in electrical contact with an object made of iron (such as a steel tank or iron pipe), electrons will be lost preferentially from the magnesium when either is confronted with an oxidizing agent. In this fashion, the magnesium protects the iron object from oxidation, thereby preventing its corrosion.

16.36 Since a 1.00 ampere current carries 6.24×10^{18} e^- in 1.00 seconds, the following equivalency holds:

$$1.00 \text{ ampere·sec} = 6.24 \times 10^{18} \text{ } e^-$$

The number of electrons carried by a 10.0 ampere current in 10.0 minutes is:

$$(10.0 \text{ min})\left(\frac{60 \text{ sec}}{1 \text{ min}}\right)\left(\frac{6.24 \times 10^{18} \text{ } e^-}{1.0 \text{ ampere·sec}}\right)(10.0 \text{ ampere}) = 3.74 \times 10^{22} \text{ } e^-$$

It takes two electrons to reduce each Cu^{2+} ion to a Cu atom, so the mass of copper the current can produce is:

$$\left(3.74 \times 10^{22} \text{ } e^-\right)\left(\frac{1 \text{ Cu atom}}{2 \text{ } e^-}\right)\left(\frac{1 \text{ mol Cu}}{6.02 \times 10^{23} \text{ Cu atom}}\right)\left(\frac{63.5 \text{ g Cu}}{1 \text{ mol Cu}}\right) = 1.97 \text{ g Cu}$$

16.37 (a) Cl = + 1; F = −1   (b) Al = + 3; H = − 1

(c) O = + 2; F = − 1   (d) H = + 1; Cl = + 1; O = − 2

(e) H = + 1; F = − 1; O = 0

CHAPTER 17

Reaction Rates And Equilibrium

17.1 When a flammable liquid achieves a high enough temperature, there is a chance that some high-energy molecules will collide with sufficient activation energy to undergo combustion. Once the reaction is initiated in this fashion, the heat that is liberated provides more energy to continue combustion in other molecules.

17.2 As the temperature is increased, so are the velocities of the molecules. This leads to an increased frequency of collision and an increase in rate.

17.3 (a) When [A] is doubled, the rate doubles. When [B] is doubled, the rate doubles.
   (b) When [A] is doubled, the rate doubles. When [B] is doubled, the rate quadruples.
   (c) When [A] is doubled, the rate doubles. When [B] is doubled, the rate does not change.

17.4 Because a catalyst is not consumed in a reaction, it is not included with the reactants. Nevertheless, to show its presence, we place it over the arrow in the equation.

17.5 The iodine color will increase. No hydrogen is introduced into the flask, so the only reaction that takes place at first is $2\ HI \rightarrow H_2 + I_2$. Because this reaction produces I_2, the intensity of the iodine color increases until equilibrium is established.

17.6 (a) $K_{eq} = \dfrac{[SO_3]^2}{[SO_2]^2[O_2]}$

   (b) $K_{eq} = \dfrac{[NCl_3]^2}{[N_2][Cl_2]^3}$

   (c) $K_{eq} = \dfrac{[NOBr]^2}{[NO]^2[Br_2]}$

   (d) $K_{eq} = \dfrac{[CO_2][H_2]}{[CO][H_2O]}$

   (e) $K_{eq} = \dfrac{[CH_4][H_2O]}{[CO][H_2]^3}$

   (f) $K_{eq} = \dfrac{[PCl_3][Cl_2]}{[PCl_5]}$

   (g) $K_{eq} = \dfrac{[O_3]^2}{[O_2]^3}$

   (h) $K_{eq} = \dfrac{[COCl_2]}{[CO][Cl_2]}$

   (i) $K_{eq} = \dfrac{[HF]^2}{[H_2][F_2]}$

17.7 (a) $K_{eq} = \dfrac{[NO]^2}{[N_2][O_2]} = \dfrac{(0.10)^2}{(1.00)(0.10)} = 0.10$

   (b) $K_{eq} = \dfrac{[I_2][Cl_2]}{[ICl]^2} = \dfrac{(0.25)(0.20)}{(0.67)^2} = 0.11$

17.7 (continued)

(c) $K_{eq} = \dfrac{[PCl_3][Cl_2]}{[PCl_5]} = \dfrac{(0.20)(0.10)}{(0.40)} = 0.050$

17.8 The equilibrium mixture will be almost entirely reactants. A very small equilibrium constant means that the relative amount of product formed is very small.

17.9 (a) shift to right   (b) shift to left   (c) shift to left   (d) shift to right
      (e) shift to left

17.10 (a) shift to right   (b) no effect   (c) shift to left   (d) shift to left

17.11 (a) shift to right   (b) no effect   (c) shift to left

17.12 (a) shift to left   (b) shift to right   (c) shift to right   (d) shift to left

17.13 $K_a = \dfrac{[H^+][F^-]}{[HF]} = \dfrac{(0.026)(0.026)}{(0.97)} = 7.0 \times 10^{-4}$

17.14 $K_a = \dfrac{[H^+][ClO^-]}{[HClO]} = \dfrac{(5.6 \times 10^{-4})(5.6 \times 10^{-4})}{(0.10)} = 3.1 \times 10^{-6}$

17.15

	HZ(aq) $\rightleftharpoons$	H$^+$(aq) +	Z$^-$(aq)
initial concentration	0.10	0	0
dissociates	−0.05	+0.05	+0.05
equilibrium concentration	0.05	0.05	0.05

$K_a = \dfrac{[H^+][Z^-]}{[HZ]} = \dfrac{(0.05)(0.05)}{(0.05)} = 5 \times 10^{-2}$

$pH = -\log(0.05) = -\log(5 \times 10^{-2})$
$\quad = -(-1.3) = 1.3$

17.16 (a) $AgBr(s) \rightleftharpoons Ag^+(aq) + Br^-(aq)$
$K_{sp} = [Ag^+][Br^-]$
$\quad = (7.1 \times 10^{-7})(7.1 \times 10^{-7}) = 5.0 \times 10^{-13}$
(b) $BaCO_3(s) \rightleftharpoons Ba^{2+}(aq) + CO_3^{2-}(aq)$
$K_{sp} = [Ba^{2+}][CO_3^{2-}]$
$\quad = (4.0 \times 10^{-5})(4.0 \times 10^{-5}) = 1.6 \times 10^{-9}$
(c) $CaF_2(s) \rightleftharpoons Ca^{2+}(aq) + 2\ F^-(aq)$
$K_{sp} = [Ca^{2+}][F^-]^2$
$\quad = (3.5 \times 10^{-4})(7.0 \times 10^{-4})^2 = 1.7 \times 10^{-10}$

17.17 (a) A reaction profile is a diagram depicting the energy changes that occur during the progress of a reaction from reactants to products.
(b) The reaction coordinate is the horizontal axis of a reaction profile and represents the progress of the reaction.
(c) The activation energy is the minimum energy required for reactants to proceed to products.
(d) The transition state is the energy maximum in the energy profile and represents the transition point between reactants and products.

17.18 (a) As the concentrations of reactants increase, the number of collisions with other reactant molecules increases, generally leading to an increased rate of reaction.
(b) As the temperature increases, the number of reacting molecules with sufficient energy to overcome the energy barrier increases, thereby leading to an increased rate of reaction .
(c) In order to undergo a successful reaction, molecules must collide with the proper orientation.

17.19 (a) The rate will be multiplied by a factor of four.
(b) The rate will be multiplied by a factor of nine.
(c) The rate will be divided by a factor of four.

17.20 (a) A catalyst provides an alternate route to products; this route has a lower activation energy.
(b) An enzyme is a biological catalyst.

17.21 (a) The rates of the forward and reverse reactions of a system at equilibrium are equal.
(b) A system at equilibrium is characterized by constant macroscopic properties.
(c) At the microscopic level, a system at equilibrium continues to undergo changes between two opposing processes.

17.22 (a) $K_{eq} = \dfrac{[I_2][Br_2]}{[IBr]^2}$ (b) $K_{eq} = \dfrac{[H_2O]^2}{[H_2]^2[O_2]}$

(c) $K_{eq} = \dfrac{[CO]^2[O_2]}{[CO_2]^2}$ (d) $K_{eq} = \dfrac{[NO]^2}{[N_2][O_2]}$

(e) $K_{eq} = \dfrac{[NO_2]^4[O_2]}{[N_2O_5]^2}$ (f) $K_{eq} = \dfrac{[CO_2]^3[H_2O]^4}{[C_3H_8][O_2]^5}$

17.23 (a) $K_{eq} = \dfrac{[CH_3OH]}{[CO][H_2]^2} = \dfrac{(1.00)}{(2.00)(1.00)^2} = 0.500$

(b) $K_{eq} = \dfrac{[CO]^2[O_2]}{[CO_2]^2} = \dfrac{(0.046)^2(0.023)}{(0.30)^2} = 5.4 \times 10^{-4}$

17.23 (continued)

(c) $K_{eq} = \dfrac{[CO_2][H_2]}{[CO][H_2O]} = \dfrac{(2.3)(2.3)}{(3.7)(3.7)} = 0.39$

17.24 If K_{eq} is large, the equilibrium mixture will be predominantly products, or the reaction will tend to go to completion. If K_{eq} is small, the equilibrium mixture will contain very little product, or the reaction will proceed very little. If K_{eq} is of intermediate magnitude, there will be comparable concentrations of reactants and products at equilibrium.

17.25 Le Chatelier's Principle: When a stress is applied to a system at equilibrium, the system will respond to relieve the stress.
(a) shift to right   (b) shift to right   (c) shift to right   (d) shift to left
(e) shift to left

17.26 (a) shift to left   (b) no effect   (c) shift to right   (d) shift to left

17.27 (a) shift to right   (b) shift to right   (c) shift to left   (d) shift to left

17.28 (a) no effect   (b) shift to left   (c) shift to right   (d) no effect
(e) shift to right   (f) shift to right

17.29 $K_a = \dfrac{[H^+][NO_2^-]}{[HNO_2]} = \dfrac{(0.0065)(0.0065)}{(0.094)} = 4.5 \times 10^{-4}$

17.30 $K_a = \dfrac{[H^+][C_7H_5O_2^-]}{[HC_7H_5O_2]} = \dfrac{(0.0036)(0.0036)}{(0.196)} = 6.6 \times 10^{-5}$

17.31

	HW	⇌	H⁺	+	W⁻
initial concentration:	1.00		0		0
dissociates:	− 0.25		+0.25		+0.25
final:	0.75		0.25		0.25

$K_a = \dfrac{[H^+][W^-]}{[HW]} = \dfrac{(0.25)(0.25)}{(0.75)} = 8.3 \times 10^{-2}$

pH $= -\log(0.25) = -(-0.60) = 0.60$

17.32 20% of 0.0140 M is: (0.20)(0.0140 M) = 0.0028 mol/liter dissociates as follows:

$$HF \rightleftharpoons H^+ + F^-$$

	HF	H$^+$	F$^-$
initial concentration:	0.0140	0	0
dissociates:	−0.0028	+0.0028	+0.0028
equilibrium concentration:	0.0112	0.0028	0.0028

$$K_a = \frac{[H^+][F^-]}{[HF]} = \frac{(0.0028)(0.0028)}{(0.0112)} = 7.0 \times 10^{-4}$$

17.33 (a) $PbCO_3 \rightleftharpoons Pb^{2+} + CO_3^{2-}$
$K_{sp} = [Pb^{2+}][CO_3^{2-}] = (3.9 \times 10^{-7})(3.9 \times 10^{-7}) = 1.5 \times 10^{-13}$
(b) $Ag_2CrO_4 \rightleftharpoons 2\ Ag^+ + CrO_4^{2-}$
$K_{sp} = [Ag^+]^2[CrO_4^{2-}] = (2 \times 7.8 \times 10^{-5})^2(7.8 \times 10^{-5}) = 1.9 \times 10^{-12}$
(c) $AgI \rightleftharpoons Ag^+ + I^-$
$K_{sp} = [Ag^+][I^-] = (9.2 \times 10^{-9})(9.2 \times 10^{-9}) = 8.5 \times 10^{-17}$

17.34 Decrease. An exothermic reaction has the heat term on the product side. Increasing the temperature would drive the equilibrium to the left. This corresponds to a decrease in K_{eq}.

17.35 Molecules escape from the liquid state to enter the vapor state, thereby creating the vapor pressure. Molecules in the vapor state return to the liquid state. At equilibrium, the rates ot these opposing processes are equal. The macroscopic properties of this system (vapor pressure) remain constant, despite continuing changes at the microscopic (molecular) level.

17.36 (a) $AgBr(s) \rightleftharpoons Ag^+(aq) + Br^-(aq)$
$[Ag^+][Br^-] = (5.0 \times 10^{-3})(3.5 \times 10^{-5}) = 1.8 \times 10^{-7}$
$1.8 \times 10^{-7} > K_{sp} = 5 \times 10^{-13}$ A precipitate will form.
(b) $BaF_2(s) \rightleftharpoons Ba^{2+}(aq) + 2\ F^-(aq)$
$[Ba^{2+}][F^-]^2 = (1.0 \times 10^{-3})(6.0 \times 10^{-4})^2 = 3.6 \times 10^{-10}$
$3.6 \times 10^{-10} < K_{sp} = 2.4 \times 10^{-5}$ A precipitate will not form.
(c) $PbBr_2(s) \rightleftharpoons Pb^{2+}(aq) + 2\ Br^-(aq)$
$[Pb^{2+}][Br^-]^2 = (3.5 \times 10^{-3})(1.0 \times 10^{-2})^2 = 3.5 \times 10^{-7}$
$3.5 \times 10^{-7} < K_{sp} = 4.6 \times 10^{-6}$ A precipitate will not form.
(d) $CuS(s) \rightleftharpoons Cu^{2+}(aq) + S^{2-}(aq)$
$[Cu^{2+}[S^{2-}] = (1.5 \times 10^{-4})(2.0 \times 10^{-5}) = 3.0 \times 10^{-9}$
$3.0 \times 10^{-9} > K_{sp} = 8 \times 10^{-36}$ A precipitate will form.

17.37 The following table will show how answers (a) through (f) are obtained.

	PCl$_5$(g)	$\rightleftharpoons$	PCl$_3$(g)	+	Cl$_2$(g)
initial concentration:	0.30		0.00		0.00
undergoes reaction:	−0.10		+0.10		+0.10
equilibrium concentration:	0.20		0.10		0.10

(a) [PCl$_5$] = 0.30 M
(b) [PCl$_3$] = [Cl$_2$] = 0.00 M
(c) 0.10 mol/L used
(d) 0.10 mol/L formed
(e) [PCl$_3$] = 0.10 M
(f) [Cl$_2$] = 0.10 M

(g) $K_{eq} = \dfrac{[PCl_3][Cl_2]}{[PCl_5]} = \dfrac{(0.10)(0.10)}{(0.20)} = 0.050$

17.38

	CO(g)	+	H$_2$O(g)	$\rightleftharpoons$	CO$_2$(g)	+	H$_2$(g)
initial concentration:	4.0		5.0		0.0		0.0
undergoes reaction:	−3.0		−3.0		+3.0		+3.0
equilibrium concentration:	1.0		2.0		3.0		3.0

$$K_{eq} = \frac{[CO_2][H_2]}{[CO][H_2O]} = \frac{(3.0)(3.0)}{(1.0)(2.0)} = 4.5$$

17.39 (a) first order   (b) third order   (c) fifth order   (d) one and one-half order
(e) first order   (f) zero order

17.40 (a) Equilibrium is a dynamic process. Although there are no changes at the macroscopic level, individual solute ions return to the solid state, while individual ions from the solid dissolve to form solute ions. Over time, some of the radioactive iodide from the solid potassium iodide enters the solution to replace solute ions that have returned to the solid.
(b) Radioactive iodide from the potassium iodide crystals enters the solution as in part a. These solute ions are now in contact with the lead(II) iodide precipitate. Since the iodide from the precipitate is also in dynamic equilibrium with the solution, radioactive iodide ions from the solution eventually replace iodide ions in the precipitate.

17.41 (a) pH = 2.37 means: $[H^+] = 4.3 \times 10^{-3}$ M

$[C_2H_3O_2^-] = 4.3 \times 10^{-3}$ M

$[HC_2H_3O_2] = 1.0$ M (dissociation is negligible)

$$K_a = \frac{[H^+][C_2H_3O_2^-]}{[HC_2H_3O_2]} = \frac{(4.3 \times 10^{-3})(4.3 \times 10^{-3})}{(1.0)} = 1.8 \times 10^{-5}$$

(b) pH = 4.23 means $[H^+] = 5.9 \times 10^{-5}$ M

$[ClO^-] = 5.9 \times 10^{-5}$ M

$[HClO] = 0.10$ M (dissociation is negligible)

$$K_a = \frac{[H^+][ClO^-]}{[HClO]} = \frac{(5.9 \times 10^{-5})(5.9 \times 10^{-5})}{(0.10)} = 3.5 \times 10^{-8}$$

CHAPTER 18

Nuclear Chemistry

18.1 (a) $^{14}_{7}\text{N}$ (b) $^{40}_{19}\text{K}$ (c) $^{90}_{38}\text{Sr}$ (d) $^{214}_{82}\text{Pb}$

      (e) $^{234}_{90}\text{Th}$ (f) $^{226}_{88}\text{Ra}$ (g) $^{239}_{94}\text{Pu}$

18.2 (a) A stream of positrons would be deflected toward the negative plate with the same amount of deflection that the beta radiation is deflected toward the positive plate.
      (b) Neutrons would not be deflected by the electrical field.

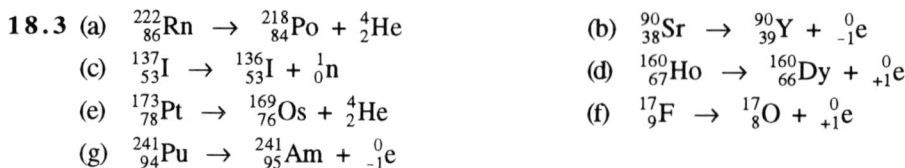

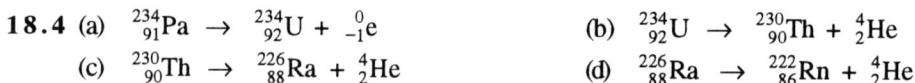

18.3 (a) $^{222}_{86}\text{Rn} \rightarrow ^{218}_{84}\text{Po} + ^{4}_{2}\text{He}$ (b) $^{90}_{38}\text{Sr} \rightarrow ^{90}_{39}\text{Y} + ^{0}_{-1}\text{e}$

      (c) $^{137}_{53}\text{I} \rightarrow ^{136}_{53}\text{I} + ^{1}_{0}\text{n}$ (d) $^{160}_{67}\text{Ho} \rightarrow ^{160}_{66}\text{Dy} + ^{0}_{+1}\text{e}$

      (e) $^{173}_{78}\text{Pt} \rightarrow ^{169}_{76}\text{Os} + ^{4}_{2}\text{He}$ (f) $^{17}_{9}\text{F} \rightarrow ^{17}_{8}\text{O} + ^{0}_{+1}\text{e}$

      (g) $^{241}_{94}\text{Pu} \rightarrow ^{241}_{95}\text{Am} + ^{0}_{-1}\text{e}$

18.4 (a) $^{234}_{91}\text{Pa} \rightarrow ^{234}_{92}\text{U} + ^{0}_{-1}\text{e}$ (b) $^{234}_{92}\text{U} \rightarrow ^{230}_{90}\text{Th} + ^{4}_{2}\text{He}$

      (c) $^{230}_{90}\text{Th} \rightarrow ^{226}_{88}\text{Ra} + ^{4}_{2}\text{He}$ (d) $^{226}_{88}\text{Ra} \rightarrow ^{222}_{86}\text{Rn} + ^{4}_{2}\text{He}$

18.5 The radiation causes ionization of the argon inside the Geiger tube. This results in a flow of electrons through the detector. The amount of current passing through the detector is a measure of the radioactivity in its vicnity.

18.6 (a) 0.50 g (one half-life)   (b) 0.75 g (two half-lives)
      (c) 72,300 years (three half-lives)

18.7 (a) $^{36}_{18}\text{Ar}$ (b) $^{10}_{5}\text{B}$ (c) $^{20}_{10}\text{Ne}$ (d) $^{76}_{33}\text{As}$

18.8 (a) The Einstein equation relates mass and its energy equivalent: $E = mc^2$.
      (b) A mass defect is the mass that is converted into energy in a nuclear reaction.
      (c) To determine the energy equivalent of matter, we multiply the mass by the square of the velocity of light. Because the velocity of light is such a large number, even small amounts of matter have a large energy equivalent.

18.9 (a) A fissionable substance is one that, upon bombardment with a neutron, undergoes fission to yield smaller product nuclei. Energy is given off in the process, as are product neutrons that bombard other nuclei to continue the chain.
      (b) Neutrons begin each fission by bombarding a fissionable nucleus. A neutron must be produced in the fission in order to sustain the chain reaction.
      (c) A chain reaction is one in which each individual reaction provides either energy or a product necessary to begin the next reaction. Thus individual reactions proceed one after another in a chain.

18.9 (continued)

(d) A critical mass is the minimum mass of a fissionable isotope that must be present in order for a chain reaction to be sustained.

(e) If there are not sufficient fissionable isotopes present in close proximity to one another, the neutrons produced in the initial reaction escape, and the chain comes to a halt.

(f) The cadmium rods moderate the reaction by absorbing excess neutrons so that the reaction does not spiral out of control.

18.10 The reactions that take place on the sun and provide us with our ultimate source of energy are fusion reactions. Without the sun's energy, life on earth would cease to exist.

18.11 24 hr = 4 half-lives: $\left(\frac{1}{2}\right)^4 = 0.0625 = 6.25\%$

48 hr = 8 half-lives: $\left(\frac{1}{2}\right)^8 = 0.00391 = 0.391\%$

18.12 Power plant safety and the storage of radioactive wastes are the primary areas of concern.

18.13 (a) $^{3}_{1}H$ (b) $^{3}_{2}He$ (c) $^{60}_{27}Co$

(d) $^{99}_{43}Tc$ (e) $^{131}_{53}I$

18.14 (a) $^{4}_{2}He$ (b) $^{0}_{-1}e$ (c) $^{0}_{+1}e$ (d) $^{1}_{0}n$ (e) $^{1}_{1}H$ or $^{1}_{1}p$

(f) A gamma ray is a high energy form of light waves.

18.15 (a) $^{185}_{79}Au \rightarrow ^{181}_{77}Ir + ^{4}_{2}He$ (b) $^{214}_{83}Bi \rightarrow ^{214}_{84}Po + ^{0}_{-1}e$

(c) $^{14}_{8}O \rightarrow ^{14}_{7}N + ^{0}_{+1}e$ (d) $^{125}_{50}Sn \rightarrow ^{125}_{51}Sb + ^{0}_{-1}e$

(e) $^{225}_{91}Pa \rightarrow ^{221}_{89}Ac + ^{4}_{2}He$ (f) $^{31}_{16}S \rightarrow ^{31}_{15}P + ^{0}_{+1}e$

18.16 (a) $^{210}_{82}Pb \rightarrow ^{210}_{83}Bi + ^{0}_{-1}e$ (b) $^{210}_{83}Bi \rightarrow ^{210}_{84}Po + ^{0}_{-1}e$

(c) $^{210}_{84}Po \rightarrow ^{206}_{82}Pb + ^{4}_{2}He$

18.17 The Geiger counter measures or "counts" the number of argon atoms ionized by the nuclear radiations.

18.18 (a) 3.0 g (one half-life)   (b) 0.15 g (three half-lives)   (c) 6400 yr (four half-lives)

18.19 (a) Transmutation is the conversion of one element to another. An <u>induced</u> transmutation is brought about by bombarding a target nucleus with a bombarding particle (such as a proton, neutron, or alpha particle).

(b) A cyclotron is used to accelerate bombarding particles to very high velocities for use in transmutation.

(c) Uranium (atomic number 92) is the naturally occurring element of highest atomic number.

18.19 (continued)

     (d) Elements above atomic number 92 have been prepared artificially by induced transmutation.

18.20 (a) $^{23}_{12}Mg$ (b) $^{256}_{101}Md$ (c) $^{241}_{95}Am$ (d) $^{257}_{103}Lr$

18.21 $E = mc^2 = \left(1.0 \times 10^{-6} \text{ kg}\right)\left(\dfrac{3.00 \times 10^8 \text{ m}}{s}\right)^2 = \dfrac{9.00 \times 10^{10} \text{ kg} \cdot \text{m}^2}{s^2}$

$= 9.00 \times 10^{10} \text{ J} = 9.00 \times 10^7 \text{ kJ}$

18.22 A breeder reactor is one in which nonfissionable isotopes capture stray neurtrons to produce new fissionable isotopes.

18.23 (a) $^{0}_{+1}e$ (b) $^{1}_{0}n$ (c) $^{2}_{1}H$ (d) $^{2}_{1}H$

18.24 (a) Tritium is the only major radioactive isotope produced during fusion, and its decay is not accompanied by any gamma radiations.

     (b) Most fuels used for fusion can be produced readily from deuterium and tritium, two readily available isotopes of hydrogen, an element that is abundant in water.

     (c) The fuels are induced to react at temperatures so high that they react with the fuel chamber. The problem of bringing about fusion before the fuel reacts with the chamber must be solved.

18.25 Alpha rays have limited penetrating powers, being unable to penetrate the skin. They are not especially dangerous externally. Beta particles have greater penetrating power and can cause severe burns. Gamma radiation has the greatest penetrating power and can cause mutation or sterility.

18.26 The likely dangers inherent in fusion reactors are not as serious as those of fission reactors. In the event of a fusion reactor failure, the reaction would quickly cool down and stop because of the high temperatures required to sustain the fusion reaction. In addition, nuclear wastes from fusion reactions are far less harmful than those from fission reactors. Fission plants, on the other hand, produce very dangerous long-lived radioactive nuclear wastes whose storage is difficult to carry out with complete safety. In the event of a reactor failure the fission reaction is capable of spiraling out of control.

18.27 (a) $^{131}_{57}La$ (b) $^{230}_{90}Th$ (c) $^{4}_{2}He$

     (d) $^{14}_{7}N$ (e) $^{0}_{-1}e$ (f) $^{1}_{1}H$ or $^{1}_{1}p$

     (g) $^{1}_{0}n$ (h) $^{146}_{57}La$

18.28 $^{241}_{95}Am \rightarrow {}^{237}_{93}Np + {}^{4}_{2}He$

18.29 $^{131}_{53}I \rightarrow {}^{131}_{54}Xe + {}^{0}_{-1}e$

18.30 $1 \text{ mol } H_2O \left(\dfrac{2 \text{ mol H atoms}}{1 \text{ mol } H_2O} \right) \left(\dfrac{6.02 \times 10^{23} \text{ atoms H}}{1 \text{ mol H atoms}} \right) \left(\dfrac{1 \text{ tritium atom}}{1.0 \times 10^{17} \text{ atoms H}} \right)$

$= 1.2 \times 10^7 \text{ tritium atoms}$

18.31 To be one-fourth the normal ratio, two half lives have elapsed; the age of the first fossil is twice 5720 years, or 11,400 years old. To be one-eighth the normal ratio, three half lives have elapsed; the age of the second fossil is three times 5720 years, or 17,200 years old.

18.32 Like sound waves, radioactivity radiates 3-dimensionally. This means that radiations spread out as they travel away from their souce, and the number of radiations entering the cross-sectional area of the Geiger tube diminishes with distance.

18.33 Radon-222 is one of the isotopes formed in the uranium-238 disintegration series. Thus, it is likely to be found in the vicinity of a uranium deposit.

18.34 Less. As the particles come together to form a stable nucleus, energy in the form of mass is lost:

$$\begin{aligned}
2 \text{ protons} + 2 \text{ neutrons} = \quad 2(1.00728 \text{ u}) + 2(1.00867 \text{ u}) \quad &= \quad 4.03190 \text{ u} \\
\underline{1 \text{ helium-4 nucleus}} \quad &= \quad \underline{4.00260 \text{ u}} \\
\text{mass lost} \quad &= \quad 0.02930 \text{ u}
\end{aligned}$$

18.35 (a) $n = -3.32 \times \log(0.10) = -3.32 \times (-1.00) = 3.32$ half-lives
(b) $n = -3.32 \times \log(0.0100) = -3.32 \times (-2.00) = 6.62$ half-lives
(c) If 99.9% has decayed, $0.1\% = 0.001$ remains:
$n = -3.32 \times \log(0.001) = -3.32 \times (-3) = 9.96$ half-lives (≈ 10 half-lives)
(d) 1.00 mg is 0.1% of 1.00 g. Thus, 9.96 half-lives must elapse.
$9.96 \times 6.0 \text{ hr} = 60 \text{ hr } (6.0 \times 10^1 \text{ hr})$
(e) $9.96 \times 24,100 \text{ yr} = 240,000 \text{ yr}$

18.36 (a) $(2 \times 2.0141 \text{ u}) - 4.0026 \text{ u} = 4.0282 \text{ u} - 4.0026 \text{ u} = 0.0256 \text{ u}$
(b) If one mole of helium-4 is produced, the loss of mass is 0.0256 g.
(c) $0.0256 \text{ g} = 2.56 \times 10^{-2} \text{ g} \left(\dfrac{1 \text{ kg}}{10^3 \text{ g}} \right) = 2.56 \times 10^{-5} \text{ kg}$
(d) $E = mc^2 = (2.56 \times 10^{-5} \text{ kg})(3.00 \times 10^8 \text{ m/sec})^2$
$= 2.30 \times 10^{12} \text{ kg·m}^2/\text{sec}^2 = 2.30 \times 10^{12} \text{ J} = 2.30 \times 10^9 \text{ kJ}$

CHAPTER 19

Organic Chemistry

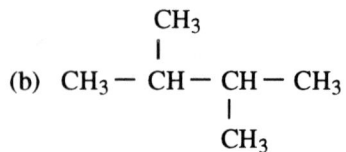

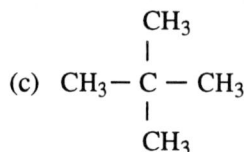

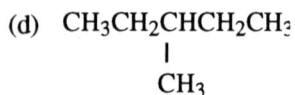

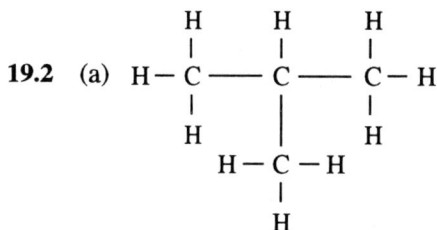

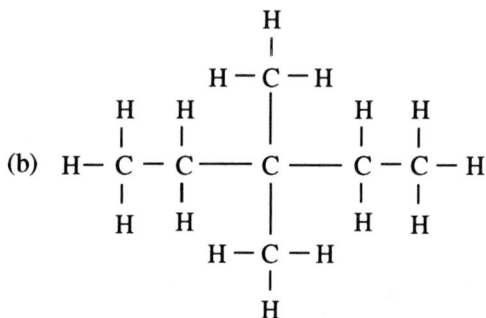

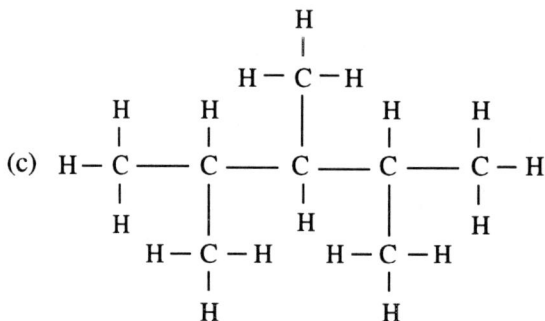

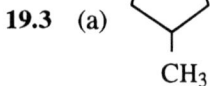

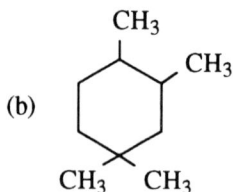

19.1 (a) $CH_3CH_2CH_3$

(b)
$$CH_3 - \overset{\overset{\displaystyle CH_3}{|}}{CH} - \underset{\underset{\displaystyle CH_3}{|}}{CH} - CH_3$$

(c)
$$CH_3 - \overset{\overset{\displaystyle CH_3}{|}}{\underset{\underset{\displaystyle CH_3}{|}}{C}} - CH_3$$

(d)
$$CH_3CH_2\underset{\underset{\displaystyle CH_3}{|}}{CH}CH_2CH_3$$

19.2 (a)

(b)

(c)

19.3 (a)

(b)

150

19.4 (a)

$$CH_3$$
$$|$$
$$CH$$
$$CH_2 \quad CH_2$$
$$| \qquad\qquad CH_3$$
$$CH_2 \quad C$$
$$CH_2 \qquad CH_3$$

(b)

$$CH_3$$
$$CH_2-CH$$
$$| \qquad CH-CH_3$$
$$CH_2$$
$$CH_2$$

19.5 $CH_3CH_2CH_2CH_2CH_2CH_3$

$$CH_3CHCH_2CH_2CH_3$$
$$|$$
$$CH_3$$

$$CH_3CH_2CHCH_2CH_3$$
$$|$$
$$CH_3$$

$$CH_3$$
$$|$$
$$CH_3CCH_2CH_3$$
$$|$$
$$CH_3$$

$$CH_3CH-CHCH_3$$
$$| \qquad |$$
$$CH_3 \quad CH_3$$

19.6 (a)

$$CH_3$$
$$|$$
$$CH_3CCH_2CH_3$$
$$|$$
$$CH_3$$

(b) $CH_3CHCH_2CH-CHCH_2CH_3$
$$\quad\;\; | \qquad\quad | \qquad\;\; |$$
$$\quad\;\; CH_3 \quad CH_3 \quad CH_3$$

(c)

$$CH_3$$
$$|$$
$$CH_3CH_2CCH_2CH_3$$
$$|$$
$$CH_2$$
$$|$$
$$CH_3$$

(d) $CH_3CH-CH-CHCH_2CH_3$
$$\qquad | \qquad | \qquad |$$
$$\qquad Br \quad Cl \quad Br$$

(e) CF_3CHCH_3
$$\qquad\quad |$$
$$\qquad\quad CH_3$$

(f) $CH_3CH_2CH-CH-CH-CHCH_2CH_2CH_3$
$$\qquad\qquad\;\; | \qquad | \qquad | \qquad |$$
$$\qquad\qquad\;\; CH_3 \quad CH_2 \quad CH_3 \quad CH$$
$$\qquad\qquad\qquad\qquad\;\; | \qquad\qquad\quad / \; \backslash$$
$$\qquad\qquad\qquad\qquad\;\; CH_3 \qquad\quad CH_3 \; CH_3$$

(g)

$$CH_3 \quad CH_3$$
$$| \qquad\quad |$$
$$CH_3C-CCH_3$$
$$| \qquad\quad |$$
$$CH_3 \quad CH_3$$

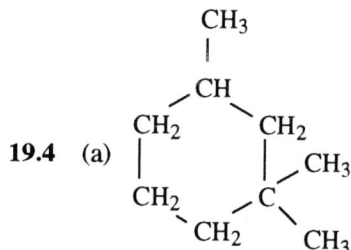

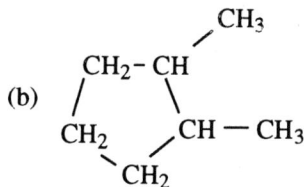

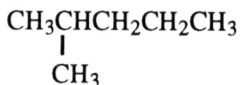

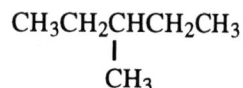

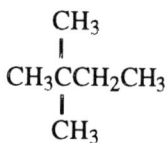

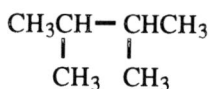

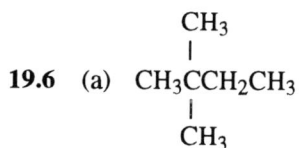

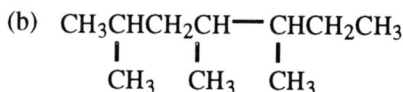

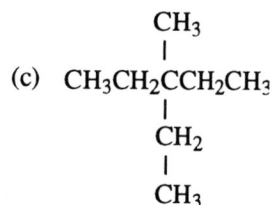

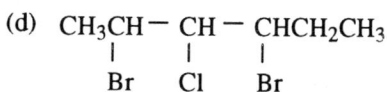

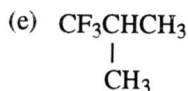

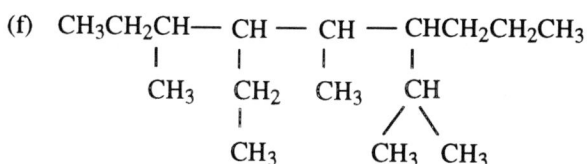

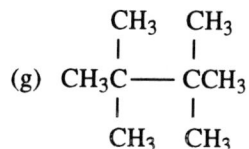

19.7 dichlorodifluoromethane

19.8 (a) 2,2-dimethylbutane
(c) 3-ethyl-3,4-dimethylhexane
(e) 2,2,3-trimethylbutane
(b) 3,5-dimethyl-5-propyloctane
(d) 2,2,3,4-tetramethylpentane

19.9   CH$_3$CH$_2$CH$_2$CH$_2$Cl   CH$_3$CHCH$_2$CH$_3$
              1-chlorobutane |
                                                  Cl

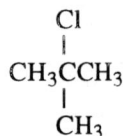

                                      2-chlorobutane

               Cl
               |
           CH$_3$CCH$_3$ CH$_3$CHCH$_2$Cl
               | |
               CH$_3$ CH$_3$

         2-chloro-2-methylpropane   1-chloro-2-methylpropane

19.10   Cl — CHCH$_2$CH$_3$ Cl — CH$_2$CHCH$_3$
                     | |
                     Cl   Cl

          1,1-dichloropropane   1,2-dichloropropane

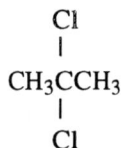

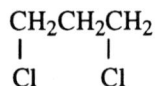

                Cl
                |
           CH$_3$CCH$_3$ CH$_2$CH$_2$CH$_2$
                | | |
                Cl   Cl   Cl

           2,2-dichloropropane   1,3-dichloropropane

19.11   CH$_2$ = CHCH$_2$CH$_2$CH$_3$ CH$_3$CH = CHCH$_2$CH$_3$
                1-pentene   2-pentene

      CH$_2$ = CCH$_2$CH$_3$ CH$_3$CH = CCH$_3$ CH$_2$ = CHCHCH$_3$
                | | |
                CH$_3$ CH$_3$ CH$_3$

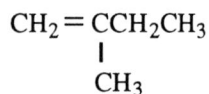

       2-methyl-1-butene   2-methyl-2-butene   3-methyl-1-butene

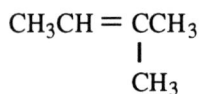

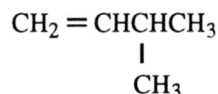

19.12   (a)   1-pentene   (b)   2-hexene   (c)3-heptene

19.13   H — C ≡ CCH$_2$CH$_2$CH$_3$ CH$_3$C ≡ CCH$_2$CH$_3$ H — C ≡ CCHCH$_3$
               1-pentyne   2-pentyne |
                                                             CH$_3$

                                                3-methyl-1-butyne

19.14   (a)   2-pentyne   (b)   1-pentyne   (c)   3-octyne

19.15 (a) $CH_3CH_2CH_2CH_2$

(b) $CH_3CH-CHCH_3$
        | |
       Br   Br

(c) $CH_3CH_2CHCH_3$
        |
       Cl

(d) $CH_3CH_2CHCH_3$
         |
        OH

19.16 (a) alkene (b) aromatic (c) alkane (d) alkyne (e) alkane

19.17 Fractional distillation is based on differences in the boiling points of the components of a mixture of hydrocarbons. As the petroleum mixture is heated, the components of the mixture with the lowest boiling points distill (or boil off) first. The mixture is separated in order of boiling points from the lowest-boiling first to the highest-boiling last.

19.18 (a) 2-butanol   (b) 1-pentanol   (c) 3-hexanol

19.19 (a) propanal   (b) 2-pentanone   (c) 3-hexanone   (d) hexanal

19.20 (a) pentanoic acid   (b) methanoic acid (formic acid)   (c) heptanoic acid

19.21 (a)
$$CH_3CH_2\overset{\displaystyle O}{\overset{\displaystyle \|}{C}}-OCH_3$$

(b)
$$CH_3-\!\!\bigcirc\!\!-\overset{\displaystyle O}{\overset{\displaystyle \|}{C}}-O-CH\overset{\nearrow CH_3}{\underset{\searrow CH_3}{}}$$

(c)
$$CH_3-\overset{\displaystyle CH_3}{\overset{\displaystyle |}{\underset{\displaystyle \underset{CH_3}{|}}{C}}}-\overset{\displaystyle O}{\overset{\displaystyle \|}{C}}-O-CH_2-\bigcirc$$

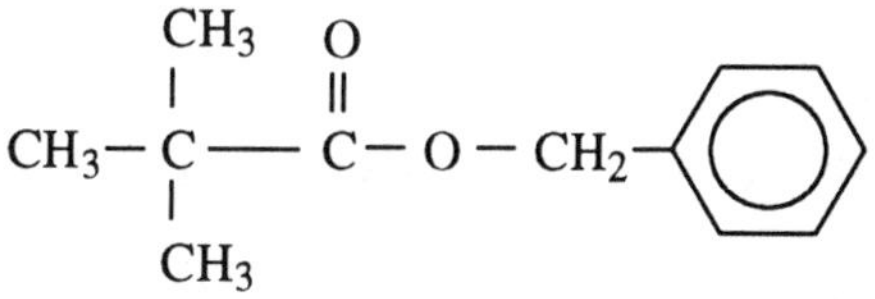

19.22 (a)
$$CH_3\overset{\displaystyle O}{\overset{\displaystyle \|}{C}}-OH \quad + \quad HO-CH_2CH_2CH_2CH_2CH_3$$

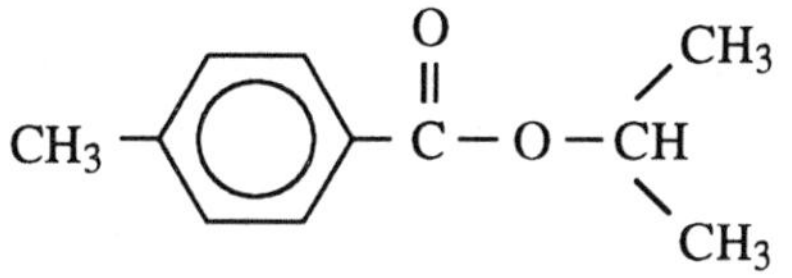

(b)
$$\bigcirc-\overset{\displaystyle O}{\overset{\displaystyle \|}{C}}-OH \quad + \quad HO-CH\overset{\nearrow CH_3}{\underset{\searrow CH_3}{}}$$

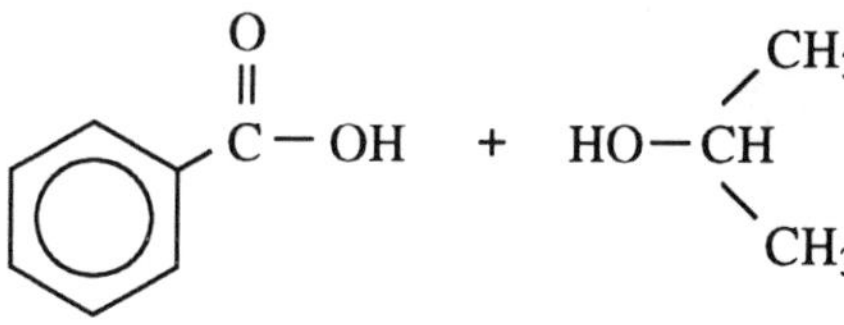

(c)
$$H-\overset{\displaystyle O}{\overset{\displaystyle \|}{C}}-OH \quad + \quad HO-CH_2CH_2CH_2CH_3$$

19.23 (a) $CH_3CH_2C(=O)-OCH_3$

(b) $CH_3CH_2CH_2CH_2CH_2C(=O)-OCH(CH_3)_2$ (isopropyl group: $-OCH$ with two CH_3 branches)

(c) Ph$-C(=O)-OCH_2CH_3$

19.24 (a) $CH_3CH_2C(=O)-NH_2$

(b) $CH_3C(=O)-NH-$Ph

19.25 (a) Ph$-C(=O)-OH$ + NH_3

(b) $CH_3C(=O)-OH$ + $H_2N-CH(CH_3)_2$

19.26 (a) $CH_3CH_2CH_2CH_2CH_2C(=O)-NH_2$ (b) $CH_3CH_2CH_2C(=O)-NH_2$ (c) $H-C(=O)-NH_2$

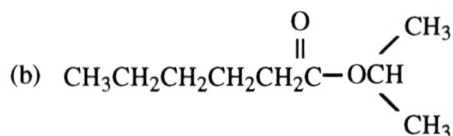

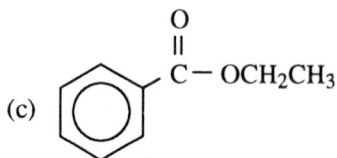

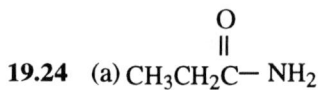

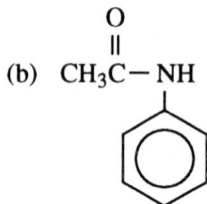

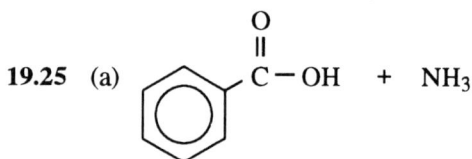

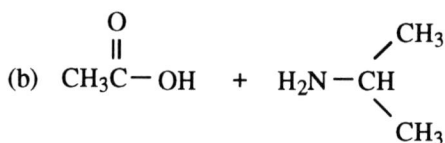

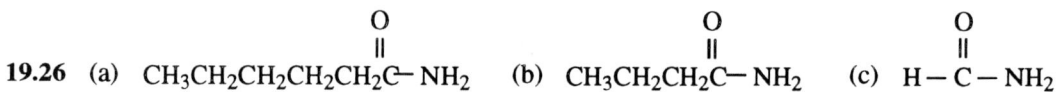

19.27 (a) alcohol, amine, phenol, (which includes an aromatic ring)
(b) carboxylic acid, amine, aromatic ring
(c) aldehyde, ether, phenol (which includes an aromatic ring)

19.28 (a) 10 (b) 3 (c) 5 (d) 11 (e) 7 (f) 9 (g) 12 (h) 2 (i) 1 (j) 6 (k) 8 (l) 4

19.29

$$\text{www} \ \overset{O}{\underset{\|}{C}} \ {\Large\{} OCH_2CH_2O - \overset{O}{\underset{\|}{C}} - \bigcirc - \overset{O}{\underset{\|}{C}} {\Large\}}_n \ O\text{www}$$

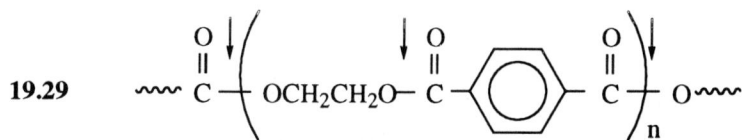

ester linkages are indicated by the arrows

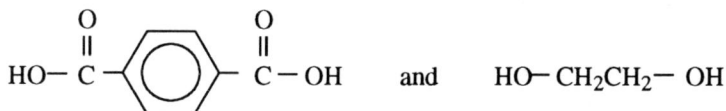

$$HO - \overset{O}{\underset{\|}{C}} - \bigcirc - \overset{O}{\underset{\|}{C}} - OH \qquad \text{and} \qquad HO - CH_2CH_2 - OH$$

component dicarboxylic acid and dialcohol

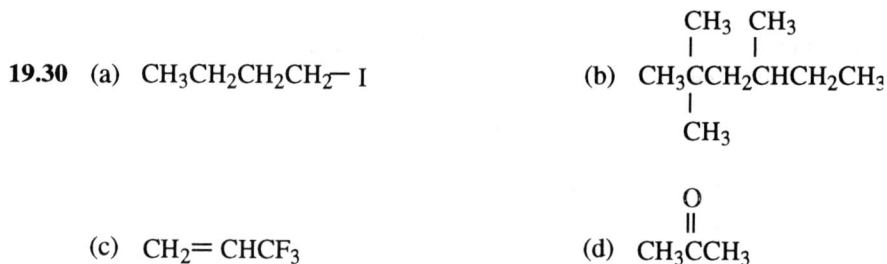

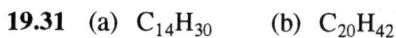

19.30 (a) $CH_3CH_2CH_2CH_2 - I$

(b) $CH_3\overset{\overset{\displaystyle CH_3}{|}}{\underset{\underset{\displaystyle CH_3}{|}}{C}}CH_2\overset{\overset{\displaystyle CH_3}{|}}{CH}CH_2CH_3$

(c) $CH_2 = CHCF_3$

(d) $CH_3\overset{O}{\underset{\|}{C}}CH_3$

19.31 (a) $C_{14}H_{30}$ (b) $C_{20}H_{42}$

19.32   *condensed*   *abbreviated*

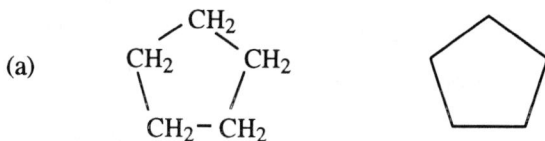

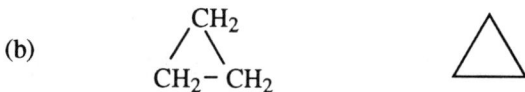

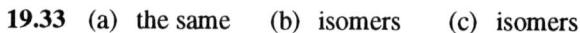

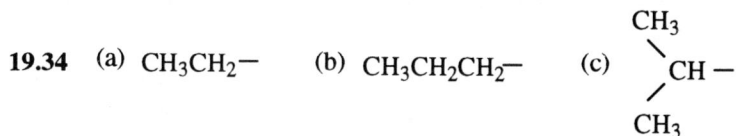

(a)

$$CH_2 \diagup \overset{CH_2}{\diagdown} CH_2$$
$$CH_2 - CH_2$$

(b)

$$CH_2 \diagup \overset{CH_2}{\diagdown}$$
$$CH_2 - CH_2$$

19.33 (a) the same   (b) isomers   (c) isomers

19.34 (a) $CH_3CH_2 -$ (b) $CH_3CH_2CH_2 -$ (c) $\overset{CH_3}{\diagdown}\underset{\diagup}{CH} - \atop CH_3$

19.35 (a) 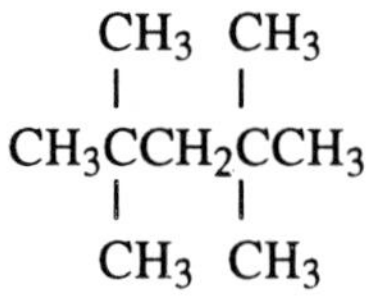

(b) 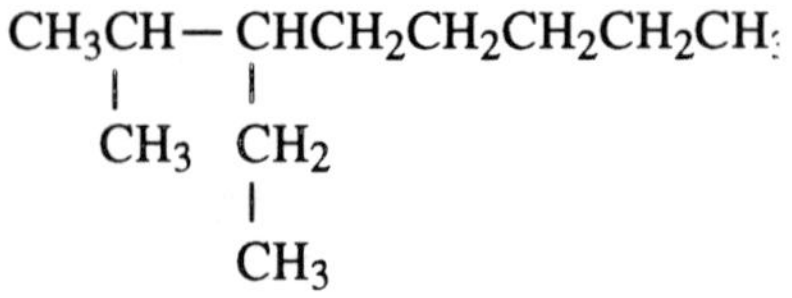

(c) 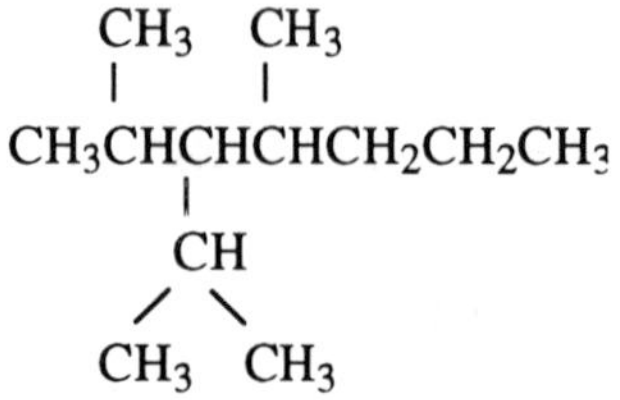

19.36 (a) 2,2-dibromopropane
(b) 2-bromo-3-chloro-4-methylhexane
(c) 4-isopropyl-3-methylheptane

19.37 $CH_3CH_2CH_2CH_2CH_2CH_2CH_3$

heptane

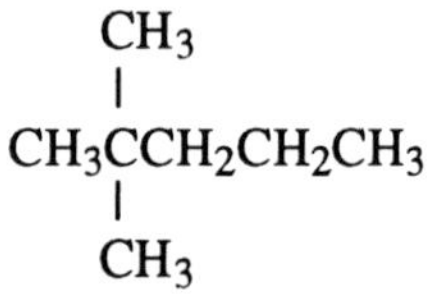

2,2-dimethylpentane

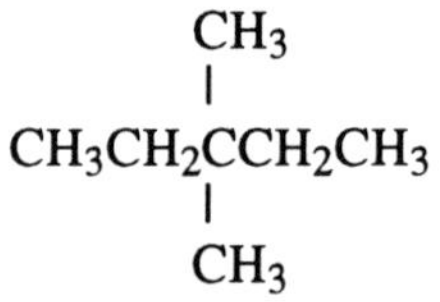

3,3-dimethylpentane

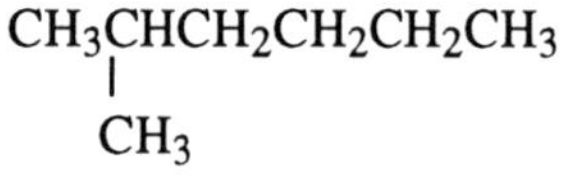

2-methylhexane

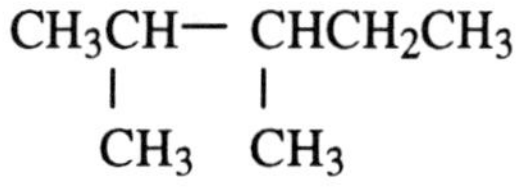

2,3-dimethylpentane

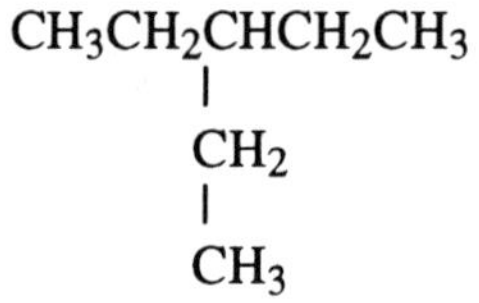

3-ethylpentane

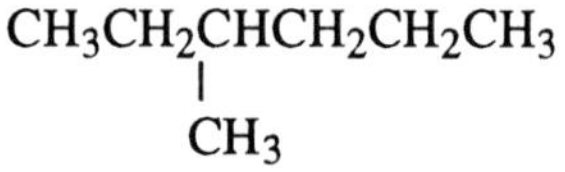

3-methylhexane

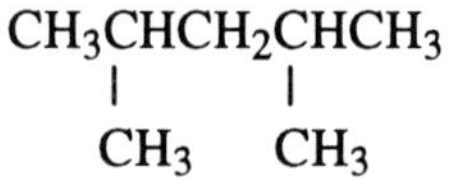

2,4-dimethylpentane

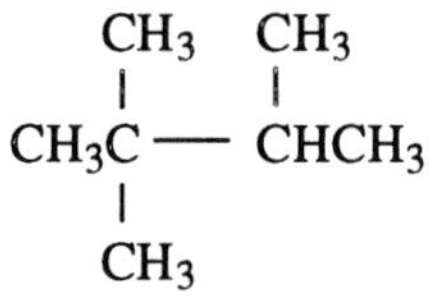

2,2,3-trimethylbutane

19.38

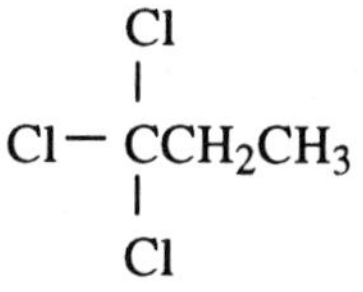

$$Cl - \underset{\underset{Cl}{|}}{\overset{\overset{Cl}{|}}{C}}CH_2CH_3$$

1,1,1-trichloropropane

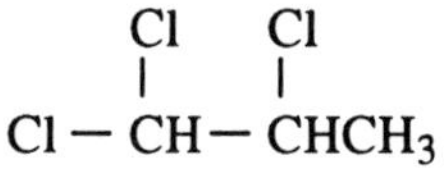

$$Cl - \underset{\underset{|}{Cl}}{\overset{\overset{Cl}{|}}{CH}} - \overset{\overset{Cl}{|}}{CH}CH_3$$

1,1,2-trichloropropane

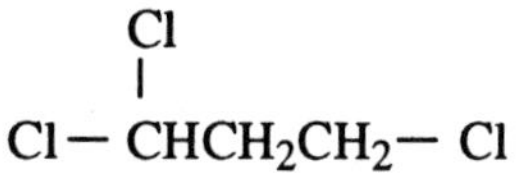

$$Cl - \overset{\overset{Cl}{|}}{CH}CH_2CH_2 - Cl$$

1,1,3-trichloropropane

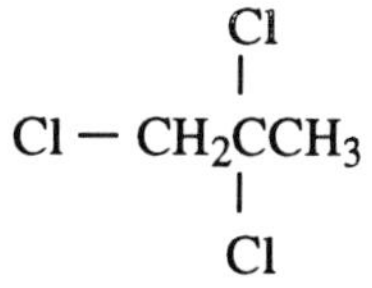

$$Cl - CH_2\underset{\underset{Cl}{|}}{\overset{\overset{Cl}{|}}{C}}CH_3$$

1,2,2-trichloropropane

$$Cl - CH_2\underset{\underset{Cl}{|}}{CH}CH_2 - Cl$$

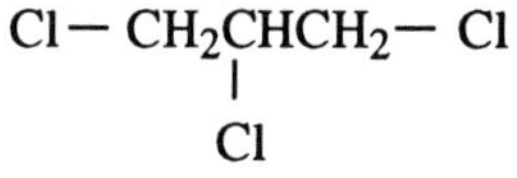

1,2,3-trichloropropane

19.39

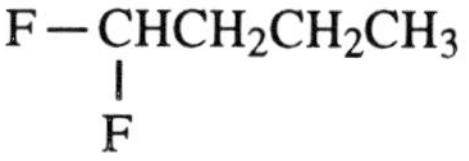

$$F - \underset{\underset{F}{|}}{CH}CH_2CH_2CH_3$$

1,1-difluorobutane

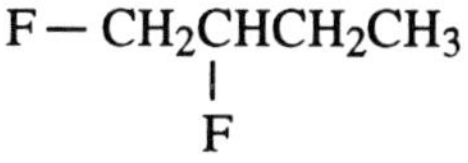

$$F - CH_2\underset{\underset{F}{|}}{CH}CH_2CH_3$$

1,2-difluorobutane

$$F - CH_2CH_2\underset{\underset{F}{|}}{CH}CH_3$$

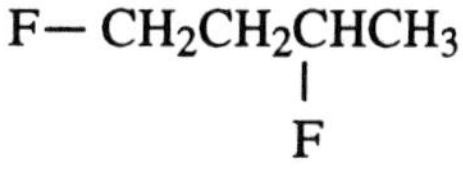

1,3-difluorobutane

$$F - CH_2CH_2CH_2CH_2 - F$$

1,4-difluorobutane

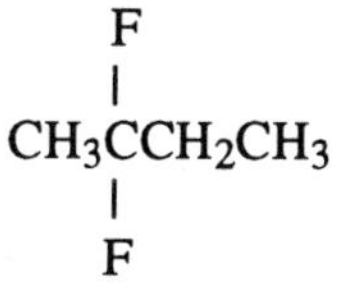

$$CH_3\underset{\underset{F}{|}}{\overset{\overset{F}{|}}{C}}CH_2CH_3$$

2,2-difluorobutante

$$CH_3\underset{\underset{F}{|}}{CH}\underset{\underset{F}{|}}{CH}CH_3$$

2,3-difluorobutane

$$F - \underset{\underset{F}{|}}{CH}\underset{\underset{CH_3}{|}}{CH}CH_3$$

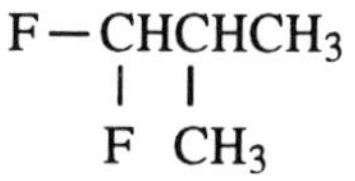

1,1-difluoro-2-methyl-
propane

$$F - CH_2\underset{\underset{CH_3}{|}}{\overset{\overset{F}{|}}{C}}CH_3$$

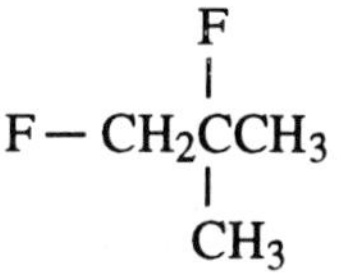

1,2-difluoro-2-methyl-
propane

$$F - CH_2\underset{\underset{CH_3}{|}}{CH}CH_2 - F$$

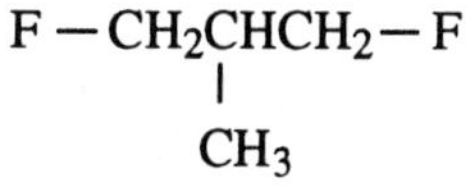

1,3-difluoro-2-methyl
propane

19.40 (a) $CH_3CH_2CH = CHCH_2CH_3$ (b) $H - C \equiv CCH_2CH_2CH_3$

(c) $CH_3\underset{\underset{CH_3}{|}}{C} = \underset{\underset{CH_3}{|}}{C}CH_3$

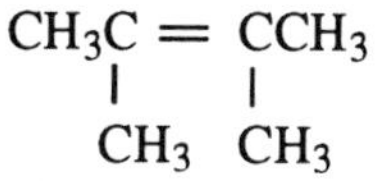

19.41 (a) 2-methyl-1-butene   (b) 4,4-dimethyl-2-pentene
      (c) 7-methyl-2-octyne   (d) 3,3-dichloro-1-butyne

19.42 (a) $CH_3CH_2CH_2CH_2CH_2CH$ (b) $CH_3CH_2CH- CHCH_2CH_3$
                                                      | |
                                                   Cl   Cl

      (c) $CH_3CH_2CH_2CHCH_2CH$ (d) $CH_3CH_2CH_2CHCH_2CH_2$
                                | |
                                Br   OH

19.43 (a) alkene (b) alkane (c) aromatic (d) alkyne

19.44 (a) C_4 – natural gas   (b) C_4 – natural gas   (c) C_8 – gasoline   (d) C_{16} – light gas oil

19.45 (a) $CH_3CH_2CHCH_2CH_2$ (b) $CH_3CHCH_2CH_2CH_2CH_2CH_2CH$
                         | |
                         OH   OH

      (c) $CH_3CH_2CH_2CHCH_2CH_2CH$
                                |
                               OH

                        O   O
                        ‖   ‖
19.46 (a) $CH_3CH_2CH_2CH_2CH_2CH_2C- H$ (b) $CH_3CCH_2CH_2CH_2CH_2CH$

                   O   O
                   ‖   ‖
      (c) $CH_3CH_2CCH_2CH_2CH_2CH$ (d) $CH_3CH_2CH_2CH_2CH_2CH_2CH_2CH_2C- H$

                                             O   O
                                           ‖   ‖
19.47 (a) $CH_3CH_2CH_2CH_2CH_2CH_2CH_2CH_2CH_2C- OH$ (b) $CH_3CH_2CH_2C- OH$

                                 O
                                 ‖
      (c) $CH_3CH_2CH_2CH_2CH_2CH_2CH_2C- OH$

                   O   O
                   ‖   ‖
19.48 (a) $CH_3CH_2CH_2C- H$ (b) $CH_3CH_2CH_2C- OH$ (c) see (a) and (b)

               butanal   butanoic acid

19.49 (a)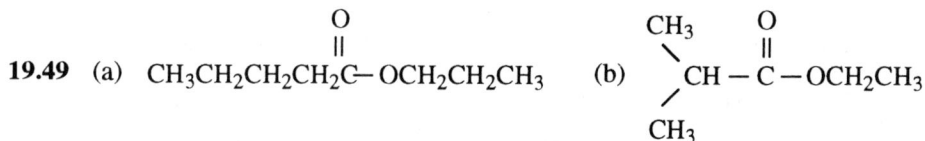

(b)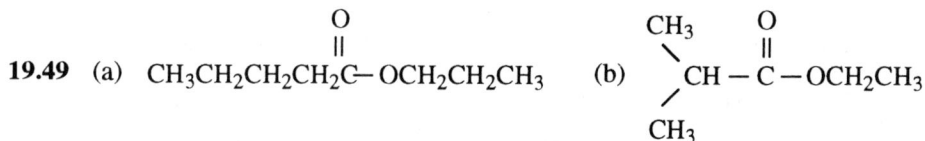

19.50 (a) 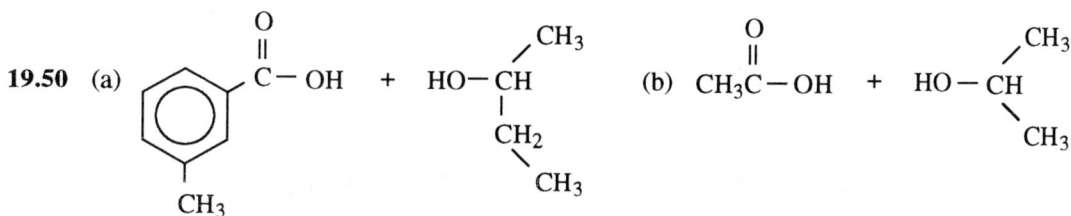   (b)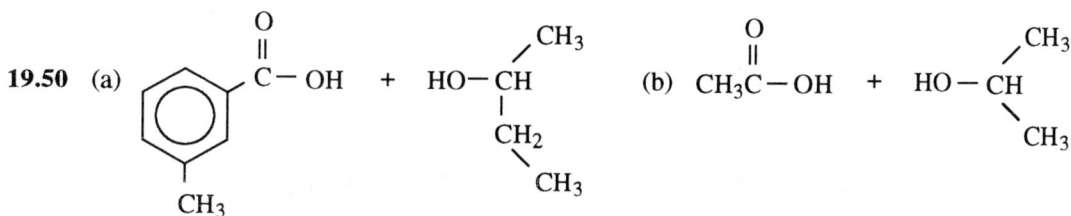

19.51 (a) 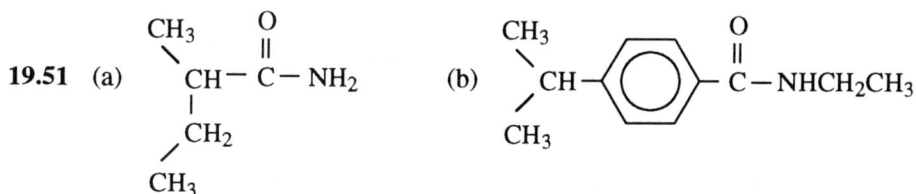   (b)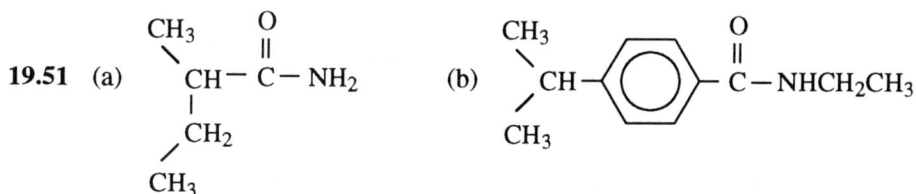

19.52 (a) 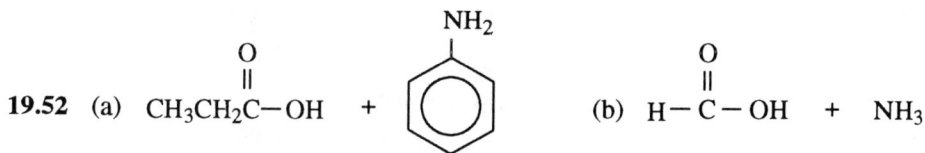   (b)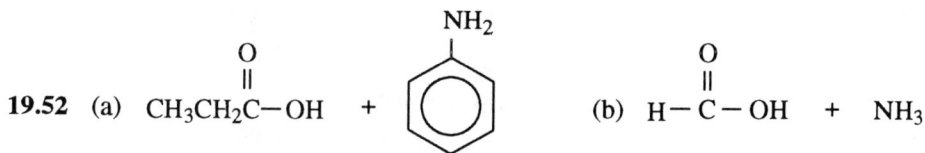

19.53 (a) aldehyde (b) amine (c) alcohol (d) amide (e) alkene
(f) halogenated alkane (g) ether (h) alkane (i) ester
(j) ketone (k) alkyne (l) carboxylic acid

15.54 Polyethylene. Individual ethene units are connected to one another in a chain:
$\left(\!\!-CH_2CH_2-\!\!\right)_n$. See table 19-6 for other examples.

15.55 (a) 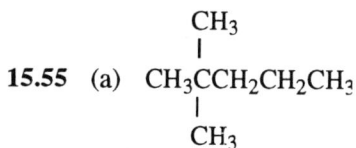   (b)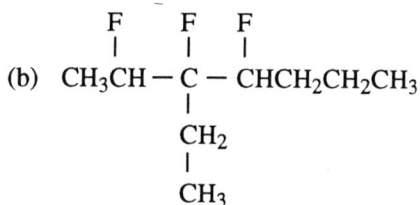

(c) 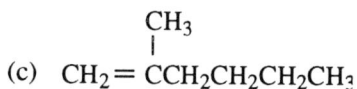   (d)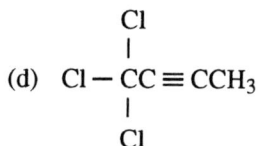

19.55 (continued)

(e) 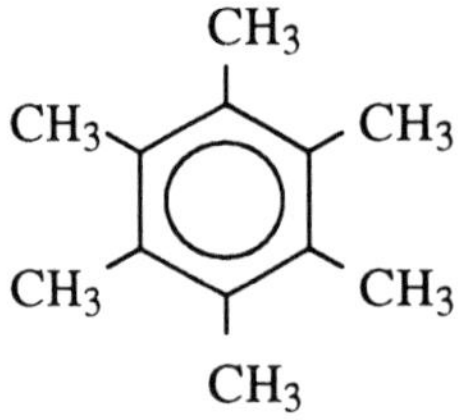 [structure: benzene ring with six CH₃ groups — hexamethylbenzene]

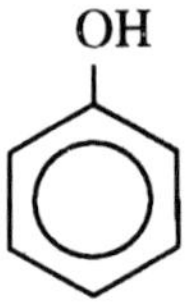

(f) 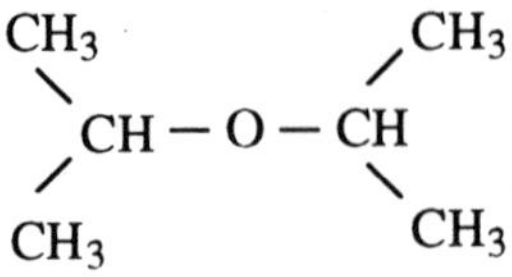 $(CH_3)_2CH-O-CH(CH_3)_2$

(g) [phenol: benzene ring with OH]

(h) $\underset{\underset{OH}{|}}{CH_3CH}CH_2CH_2CH_2CH_2CH_2CH$

(i) $CH_3CH_2\overset{\overset{O}{||}}{C}CH_2CH_3$

(j) $CH_3CH_2\overset{\overset{O}{||}}{C}-H$

(k) $CH_3CH_2CH_2CH_2CH_2CH_2CH_2CH_2\overset{\overset{O}{||}}{C}-OH$

(l) $CH_3CH_2CH_2CH_2\overset{\overset{O}{||}}{C}-OCH_2CH_2CH_2CH_3$

(m) $CH_3CH_2\overset{\overset{O}{||}}{C}-NH_2$

19.56 (a) 2,2-dimethylpropane   (b) 3,3,5-trimethyloctane
        (c) 4-ethyl-3,5-dimethylheptane   (d) 3-methyl-1-pentene
        (e) 2-heptyne   (f) 3-nonanol
        (g) 3-nonanone   (h) octanal
        (i) hexanoic acid   (j) isopropyl hexanoate
        (k) pentanamide

19.57 (b)  [methylcyclopropane structure]
     (c) 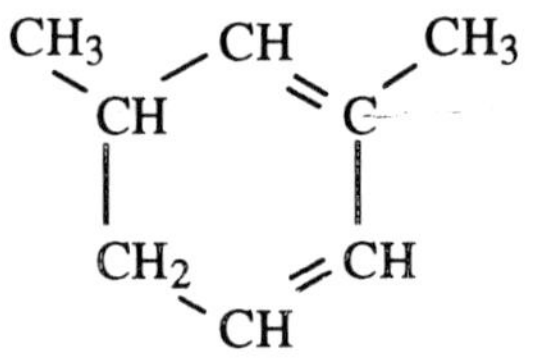 [substituted cyclohexadiene structure]
     (d) 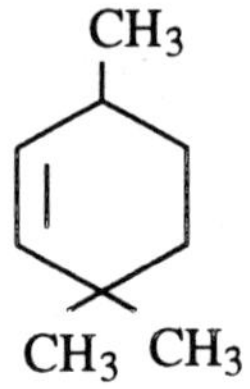 [1,4,4-trimethylcyclohexene structure]
     (e) 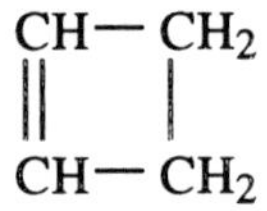 [cyclobutene structure]

19.58 $H_2N-CH_2\overset{\overset{O}{||}}{C}-NH-CH_2\overset{\overset{O}{||}}{C}-OH$

19.59 (a) 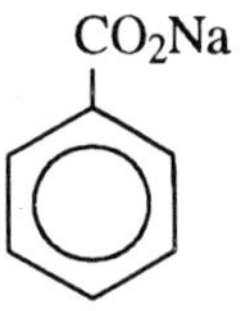 [benzene ring with CO₂Na]
     (b) 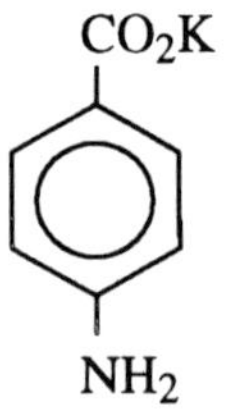 [benzene ring with CO₂K and NH₂]
     (c) $[CH_3(CH_2)_{16}CO_2]_2Ca$

     (d) $CH_3CH_2CH_2CO_2Na$ (e) $CO_2 + H_2O$

19.60 (a)

Cl, Cl, Cl, Cl, Cl, OH (pentachlorophenol) + NaOH → (O₂N, NO₂, NO₂, OK) + H_2O

(b)

(O₂N, OH, NO₂, NO₂) + KOH → (O₂N, ONa, NO₂, NO₂) + H_2O

(c)

(OH, OH) + 2 NaOH → (ONa, ONa) + H_2O

19.61 (a) $CH_3 - \overset{..}{\underset{..}{O}} - H ------ \overset{..}{\underset{}{O}} - CH_3$
                                                                        |
                                                                        H

(b) $CH_3 - \overset{..}{\underset{..}{O}} - H ------ \overset{..}{\underset{}{O}} - CH_2CH_3$
                                                                        |
                                                                        CH_2
                                                                        |
                                                                        CH_3

(c) $CH_3 - \overset{..}{\underset{..}{O}} - H ------ \overset{}{\underset{}{N}} - CH_3$
                                                        | |
                                                       (CH₃ above) CH_3
                                                        |
                                                        CH_3

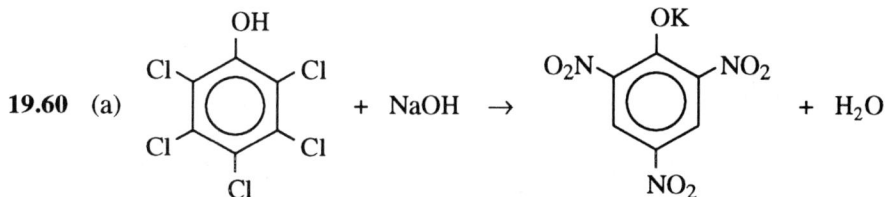

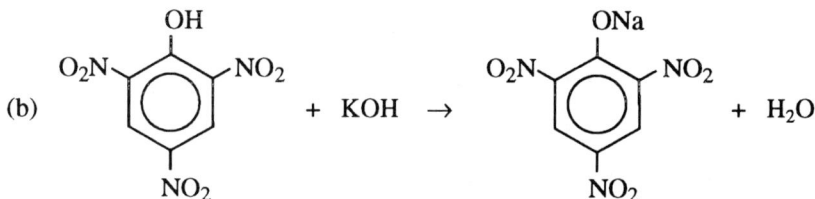

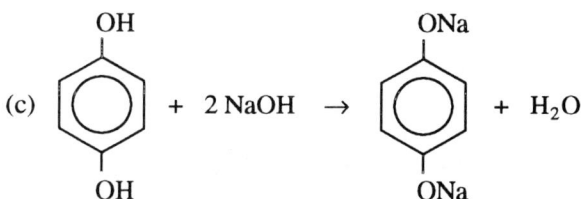

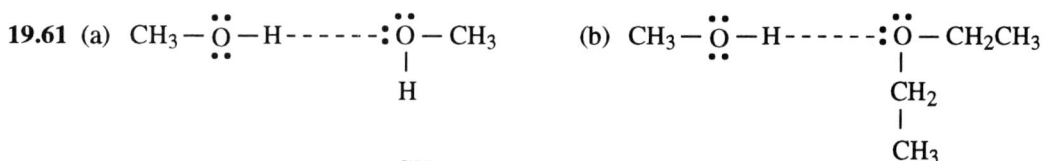

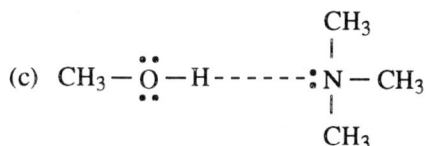

19.62 (a) alcohols, amides of the type RCONH₂, amines of the type RNH₂, carboxylic acids, and phenols
(b) aldehydes, amines of the type R₃N, esters, ethers, and ketones
(c) alkanes, alkenes, alkynes, and halogenated hydrocarbons

19.63 Dissolve the mixture in ether and shake with aqueous sodium hydroxide. Hexanoic acid will react with sodium hydroxide, forming a salt which dissolves in the water. 1-Hexanol does not react with sodium hydroxide and will remain in the ether. Separation of the immiscible liquid layers results in separation of the components of the mixture.

19.64 Rinse the flask with acetone, which dissolves any organic compounds present. Follow this with a water rinse, which removes the acetone. Wash the flask with soap and water, followed by a final rinse with distilled water.

CHAPTER 20

Biochemistry

20.1 Organisms need food to supply raw materials for the repair of damaged parts, to provide materials for reproduction and growth, and to provide energy to carry on the work of the cell.

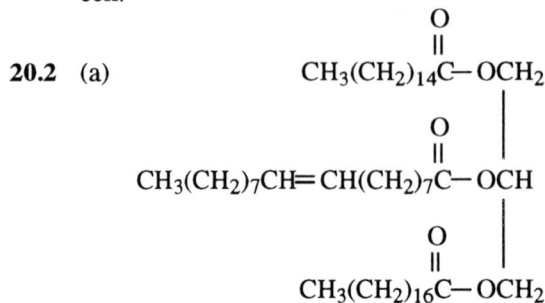

20.2 (a)

$$CH_3(CH_2)_{14}\overset{\displaystyle O}{\overset{\|}{C}}-OCH_2$$

$$CH_3(CH_2)_7CH=CH(CH_2)_7\overset{\displaystyle O}{\overset{\|}{C}}-OCH$$

$$CH_3(CH_2)_{16}\overset{\displaystyle O}{\overset{\|}{C}}-OCH_2$$

(b) Animal. This fat is largely saturated.

20.3 $CH_3(CH_2)_{14}COOH + NaOH \rightarrow CH_3(CH_2)_{14}COO^-Na^+ + H_2O$

The product of this reaction (sodium palmitate) could be used as a soap.

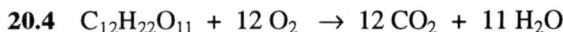

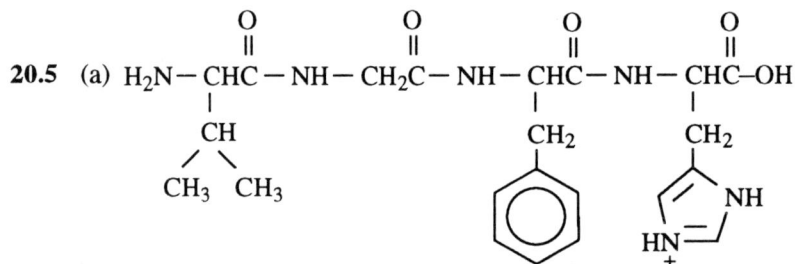

20.4 $C_{12}H_{22}O_{11} + 12\,O_2 \rightarrow 12\,CO_2 + 11\,H_2O$

20.5 (a) $H_2N-\overset{\displaystyle}{C}HC\overset{\displaystyle O}{\overset{\|}{}}-NH-CH_2C\overset{\displaystyle O}{\overset{\|}{}}-NH-\overset{\displaystyle}{C}HC\overset{\displaystyle O}{\overset{\|}{}}-NH-\overset{\displaystyle}{C}HC\overset{\displaystyle O}{\overset{\|}{}}-OH$

with side chains: $CH(CH_3)(CH_3)$ isopropyl group on first residue; CH_2–phenyl on third residue; CH_2–imidazolium ring (with NH, HN^+) on fourth residue.

20.6 The enzymes capable of breaking down starch into individual glucose units will not break down cellulose, because the specific geometry around the carbon connecting the two sugars does not permit cellulose to fit properly into the enzyme. In a similar fashion, starch molecules will not fit properly into the enzyme that breaks down cellulose. Consequently, the enzymes that break down cellulose are not capable of breaking down starch.

20.7 (a) Hydrophobic interactions are attractive interactions between nonpolar side chains.
(b) Salt-type interactions involve ionic attractions between oppositely-charged atoms on the side chains of two amino acid residues.
(c) Disulfide linkages are covalent sulfur-sulfur bonds between the side chains of cysteine residues on two amino acid residues.

20.8 A vitamin A molecule has alkene groups (carbon-carbon double bonds) and an alcohol functional group. A vitamin E molecule has a phenolic group (which includes an aromatic ring), and an ether group.

20.9 m-RNA: 3' A-A-U-A-G-A-C-G-U-C-A-U-U-C-C-G-U-A   5'
               5' A-U-G-C-C-U-U-A-C-U-G-C-A-G-A-U-A-A   3'

polypeptide: met - pro - tyr - cys - arg - stop

20.10 Adenosine monophosphate and diphosphate:

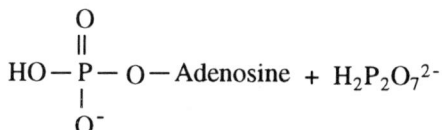

$$HO-\overset{\overset{\displaystyle O}{\|}}{\underset{\underset{\displaystyle O^-}{|}}{P}}-O-Adenosine \ + \ H_2P_2O_7^{2-}$$

20.11 Segments of DNA which are responsible for the hearty skins in crops other than tomatoes are cleaved out of their native crops and spliced into the DNA of tomatoes. This leads to hearty skins in the tomato crop.

20.12 Adrenaline contains a phenol group (which includes an aromatic ring), an alcohol group, and an amino group.

20.13 Aspirin includes a carboxylic acid group and an ester group.

20.14 Energy comes from the sun, which supplies energy for photosynthesis in the plant kingdom. Grazing animals eat plants, harnessing the energy stored from the photosynthesis process. Carnivorous animals then eat the grazing animals, harnessing the energy released from the metabolism of the animal food sources.

20.15 A cell takes in raw materials (food) which are broken down into smaller building blocks for repair, growth, and reproduction. In the process, energy may be liberated or stored in ATP, to be used later to supply energy for the cell's work. In the course of these chemical processes, wastes are produced which must be removed from the cell.

20.16. (a) A lipid is a water-insoluble component of a cell which is soluble in nonpolar solvents.
(b) A fat is a triester of glycerol (a trialcohol) and three long-chain carboxylic acids, known as *fatty acids*.
(c) A fatty acid is a long-chain carboxylic acid, usually 14 to 22 carbon atoms long.
(d) A saturated fat is one in which the fatty acids composing the triester lack any carbon-carbon double bonds. An unsaturated fat is one with carbon-carbon double bonds.
(e) Animal fats tend to have relatively few double bonds. In contrast, vegetable fats tend to be relatively high in unsaturation. Saturated fats have been linked to elevated cholesterol levels and increased risk of heart disease. Unsaturated fats are preferred from a health standpoint.

20.17 A soap molecule contains a nonpolar portion and a polar portion. A group of soap molecules are capable of dissolving an oil or grease droplet by surrounding it with their nonpolar portions. Generally, the soap molecules form a spherical arrangement with the nonpolar ends pointed inward, where they surround the oil droplet. The polar portions point away from the oil droplet, forming a layer of polar groups at the surface of the sphere, where they can dissolve in water, a polar solvent. In this fashion, the soap molecules render otherwise insoluble grease or oil molecules soluble in water.

20.18 (a) A carbohydrate is a covalently bonded organic molecule, containing an aldehyde or ketone group and two or more alcohol groups.
  (b) Carbohydrates are a major source of energy.
  (c)

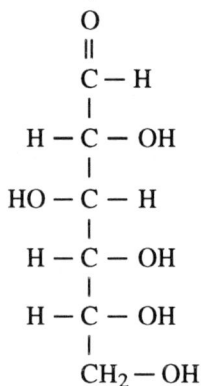

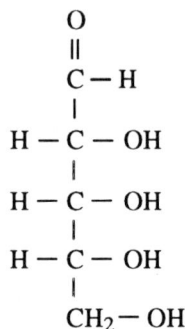

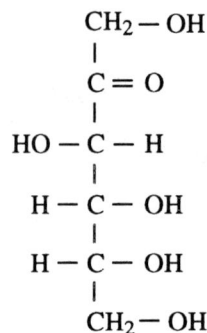

| glucose | ribose | fructose |

  (d) sucrose and lactose
  (e) starch and cellulose

20.19 Enantiomers are molecules which are mirror images that are not superimposable. D-glyceraldehyde and L-glyceraldehyde are enantiomers.

20.20 The geometry of the carbon bearing the oxygen that holds the glucose units together is different in these two polysaccharides. As a result of that small difference, cellulose will not fit into the enzyme humans use to break down starch into its component glucose units. Consequently, humans cannot harness the energy potential present in cellulose.

20.21 (a) Proteins supply structural materials that make up bones, muscles, skin, and other tissues. Many proteins act as enzymes, which are biochemical catalysts that permit the many chemical reactions that comprise the life function to be carried out in a fashion that permits living organisms to sustain the life process.
  (b) Proteins are built from amino acids.
  (c) Essential amino acids are those amino acids that must be supplied through the diet, because our bodies are not capable of biosynthesizing them.

20.22

$$\text{H}_2\text{N}-\underset{\underset{\underset{\underset{\underset{\text{CH}_3}{|}}{\text{S}}}{\overset{|}{\text{CH}_2}}}{\overset{|}{\text{CH}_2}}}{\overset{\overset{\displaystyle\text{O}}{\overset{\|}{}}}{\text{CHC}}}-\text{NH}-\underset{\underset{\text{CH}_3}{|}}{\overset{|}{\underset{\text{CH}-\text{OH}}{}}}\overset{\overset{\displaystyle\text{O}}{\|}}{\text{CHC}}-\text{NH}-\overset{\overset{\displaystyle\text{O}}{\|}}{\text{CHC}}-\text{NH}-\overset{\overset{\displaystyle\text{O}}{\|}}{\text{CHCO}_2\text{H}}$$

(chain with side chains: Met–Thr–Ile–Tyr residues; third residue side chain CH–CH₃ / CH₂ / CH₃; fourth residue side chain CH₂–C₆H₄–OH)

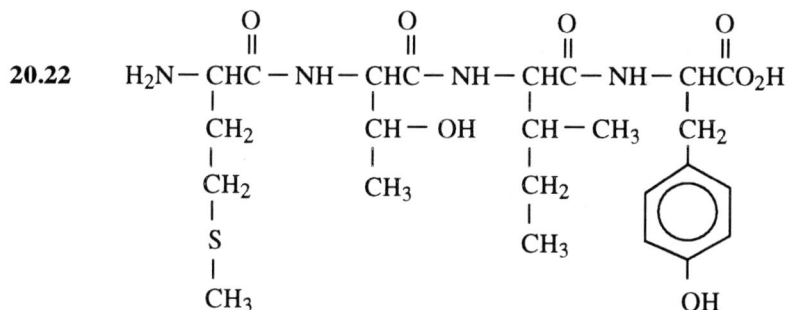

20.23   Ser-Ala-Cys

20.24 (a) The *primary structure* is the amino acid sequence of the protein chain. *Secondary structure* refers to the presence of any α-helix or any β-pleated sheet. *Tertiary structure* is the three-dimensional manner in which the chain folds up. *Quaternary structure* refers to the structure of two or more individual protein units which work together as a unit.

(b) The α-helix is a coil made by the protein chain. Carbonyl oxygens and amide hydrogens that lie on adjacent loops of the coil hydrogen bond to one another, holding the chain in this coiled arrangement. The β-pleated sheet is an arrangement of neighboring protein chains which are stretched out in a zig-zag fashion. In this case, carbonyl oxygens and amide hydrogens from neighboring chains hold the chains together in this elongated fashion.

(c) A change in an amino acid changes the side-chain normally present in that position of the protein chain. This can alter the interactions that hold various side chains together in the three-dimensional structure, thereby altering the shape of the overall three-dimensional structure assumed by the protein.

20.25 (a) A coenzyme is a substance that must be present for an enzyme to function.

(b) Many vitamins function as coenzymes.

(c) When a vitamin is lacking in the diet, the enzyme for which it is a coenzyme will not function properly. Thus, some important chemical reaction required for normal biological function is not carried out, and this translates into illness.

(d) Night blindness (A), beriberi (B_1, thiamine), and scurvy (C) are three diseases caused by vitamin deficiencies.

20.26 (a) DNA, which makes up the *genes*, is responsible for the regulation of the cell.

(b) mRNA is an intermediate in the process of taking information from the DNA and converting it into a protein, which is often an enzyme.

(c) A mononucleotide is the smallest unit of a DNA or RNA strand. It is composed of a ribose or deoxyribose sugar, a phosphate group, and a pyrimidine or purine base.

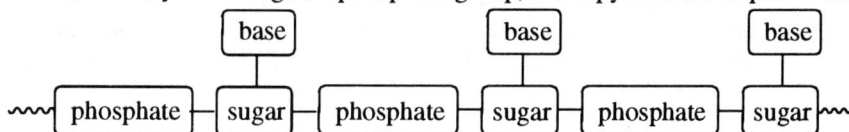

(d)

```
              base              base              base
               |                 |                 |
~~| phosphate |-| sugar |-| phosphate |-| sugar |-| phosphate |-| sugar |~~
```

20.26 (continued)

    (e) DNA is composed of two strands which are intertwined in a helical arrangement known as the *double helix*. The two strands are held together by hydrogen bonding between *complementary* base pairs on the neighboring strands.

20.27

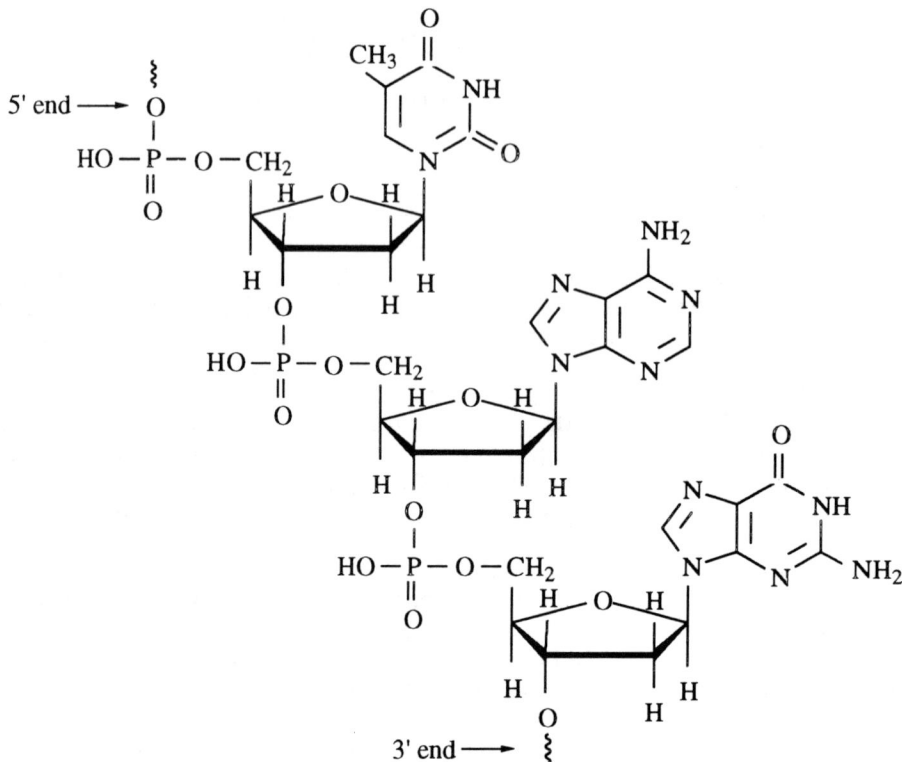

20.28 (a) DNA is a double stranded polynucleotide, in which the two strands are held together in a double helix. RNA is a single stranded polynucleotide. Strands of both DNA and RNA consist of a backbone composed of alternating phosphate and sugar units, with a pyrimidine or purine base attached to each sugar. In DNA, the sugar is deoxyribose, whereas in RNA it is ribose. In DNA the four bases are adenine, guanine, cytosine, and thymine. In RNA, the bases are adenine, guanine, cytosine, and uracil.

    (b) The bases in DNA come in pairs, known as *complementary bases*. The pairs are adenine-thymine and cytosine-guanine. When a cytosine base appears on one strand, it is held via hydrogen bonding to a guanine base on the neighboring strand. Similarly, an adenine base on one strand is held to thymine on the neighboring strand. When a strand replicates, it acts as a template for the complementary strand, by selecting for the proper complementary base. During replication, the strands unravel, and each strand produces its complementary strand, resulting in two new double helixes, identical to the original.

20.28 (continued)
- (c) When a DNA strand is transcibed to mRNA, the process is similar to DNA replication. Each base on the portion of DNA being transcribed selects its complementary base, except that adenine selects uracil, rather than thymine. Thus, the mRNA produced contains adenine, cytosine, guanine, and uracil.
- (d) Each three-base sequence in mRNA codes for one of the amino acids. These three-base sequences are known as codons. (There is generally more than one codon for each of the amino acids.) As the mRNA is "read," the amino acids are attached in the order defined by the codons in the mRNA, thereby leading to a primary structure defined by the mRNA.

20.29 *Replication* is the reproduction of a complementary DNA strand by one of the strands. When replication occurs, the strands unravel, and each strand "replicates" its complement, resulting in two new double helixes, identical to the original. *Transcription* is the process of producing a mRNA molecule from a portion of a DNA strand. This process is similar to replication, except that adenine selects uracil rather than thymine. In addition, the mononucleotide units selected contain a ribose sugar, rather than a deoxyribose sugar. *Translation* is the process of producing a protein from a mRNA segment. In this process, three-base sequences, known as "codons," are the code for one of the amino acids. As the mRNA is read, amino acids are attached in the order of the codons contained in the mRNA.

20.30 *Complementary* refers to either a (pyrimidine or purine) base or a polynucleotide strand. In DNA, the bases are paired, such that adenine is always adjacent to thymine in the neighboring strand and cytosine is always adjacent to guanine. Thus, either the bases or the strands are said to be complementary. In transcription to mRNA the same principle holds, except that uracil in mRNA is the complement of adenine in the DNA being transcribed. The term *template* is used to describe the role of a DNA strand in producing its complementary strand. Since each base always pairs with its complementary base, the strand acts as a template, selecting the order in which the bases on the complementary strand are assembled. Polynucleotide strands have direction, generally written and read from the 5' end to the 3' end. When the two strands of a DNA molecule are assembled, the strands run in opposite directions. This arrangement is said to be *antiparallel.*

20.31 When ATP releases energy, one of the phosphate linkages in the triphosphate bond is broken as follows:

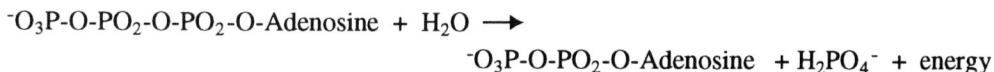

$^-O_3P\text{-}O\text{-}PO_2\text{-}O\text{-}PO_2\text{-}O\text{-Adenosine} + H_2O \longrightarrow$

$^-O_3P\text{-}O\text{-}PO_2\text{-}O\text{-Adenosine} + H_2PO_4^- + energy$

Energy is released in the process.

20.32 Gene splicing is the process of removing a gene from one cell and inserting it into the DNA of another cell. Many substances that are difficult to produce in the laboratory are produced by cellular reactions. By identifying the segment of DNA that is responsible for producing the enzymes that lead to the chemical, the segment may be "spliced" into the DNA of a bacterial cell, so that the bacteria produces the desired substance. In this fashion, a colony of bacteria can be developed and induced to produce the desired substance.

20.33 (a) A hormone is a substance manufactured by one of the endocrine glands. Hormones serve powerful regulatory functions.

    (b) adrenaline, insulin, testosterone, and progesterone are hormones.

20.34 (a)

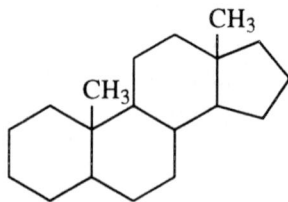

    (b) A cholesterol molecule has an alcohol group, a carbon-carbon double bond, two methyl groups, and an octyl group that is characteristic of steroids.

    (c) Testosterone has a ketone group, a carbon-carbon double bond, two methyl groups, and an alcohol group.

    (d) Progesterone is identical to testosterone, except that it has a ketone-containing group instead of the alcohol group.

20.35 The birth control pill contains a ketone group, an alkene group (carbon-carbon double bond), an alkyne group (carbon-carbon triple bond), and an alcohol group.

20.36 There are many over-the-counter medications. As an example, a bottle of generic Ibuprofen in the home of the author contained the following information:

    (a) The expected effects are relief of pain and fever, as experienced from the common cold, headaches, toothache, backache, arthritis, and menstrual cramps.

    (b) Among the warnings on a bottle is a caution to pregnant women not to take this product during the last three months of pregnancy unless directed to do so by a physician. The product may cause problems in the unborn child or complications during delivery.

20.37 Vitamins must be obtained through the diet, because they are not produced internally. They act as coenzymes, being essential for the proper functioning of important enzymes. Hormones are produced in the body by one of the endocrine glands. These chemicals have profound physiological regulatory effects. Drugs are chemicals that are introduced into the body. They may act via a wide range of mechansims. The intended function of drugs is to combat disease by relieving either the cause or symptoms of the disease.

20.38 $CH_3CH_2CH_2CH_2CH_2CH_2CH_2CH_2$ $CH_2CH_2CH_2CH_2CH_2CH_2CH_2CO_2]$

$$\underset{H}{}C = C \underset{}{}$$

cis-oleic acid

$CH_3CH_2CH_2CH_2CH_2CH_2CH_2CH_2$ H

$$C = C$$

H $CH_2CH_2CH_2CH_2CH_2CH_2CH_2CO_2]$

trans-oleic acid

20.39 The molecular mass of each glucose unit is 180.0 ($C_6H_{12}O_6$) The mass of exactly 2400 glucose units is $2400 \times 180.0 = 432,000.0$. However, every time two sugar molecules connect, a molecule of water is lost. For 2400 glucose units, that requires 2399 connections (for example, 2 sugars need only 1 connection); or a mass of $2399 \times 18.0 = 43,182$ is lost. Thus, the mass of the cellulose molecule described is $432,000.0 - 43,182.0 = 388,818.0$.

20.40 Glycine has the formula $H_2NCH_2CO_2H$ (or $C_2H_5O_2N$). Its molecular mass is 75.0. Therefore, 100 glycine units would have a molecular mass of 7500.0. However, every time two amino acids connect to form a peptide bond, a molecule of water is lost. For 100 glycine units that requires 99 connections (for example, 2 amino acids need only 1 connection); or a mass of $99 \times 18.0 = 1782.0$ is lost. Thus, the mass of the protein in question is $7500.0 - 1782.0 = 5718.0$.

20.41 Reactions occur when reacting molecules collide. Enzymes hold the reacting molecules in positions adjacent to one another so that the reactions between them occur very efficiently. This permits the reactions to occur at rates enormously faster than they would in the absence of the catalyst.

20.42

$$CH_3(CH_2)_{16}\overset{\overset{O}{\|}}{C}-OCH_2$$

$$CH_3(CH_2)_7CH = CH(CH_2)_7\overset{\overset{O}{\|}}{C}-OCH$$

$$HO-\overset{\overset{O}{\|}}{\underset{\underset{OH}{|}}{P}}-OCH_2$$

20.43 Ala-Cys-Ala-Val-Val-His-Cys-Phe-Ala-Val

20.44 (a) Lys-Gly-Leu-Ala
       (b) Lys-Ala-Trp-Pro

20.45 (a) mRNA: 3' A-U-U-A-A-G-G-C-A-C-A-U 5'
            mRNA: 5' U-A-C-A-C-G-G-A-A-U-U-A 3'
            peptide: Tyr - Thr - Glu - Leu
       (b) mRNA: 3' A-U-U-C-A-G-G-C-A-C-A-U 5'
            mRNA: 5' U-A-C-A-C-G-G-A-C-U-U-A 3'
            peptide: Tyr - Thr - Asp - Leu

20.46 Organic compounds tend to be nonpolar. Thus, the cavity of the enzyme is nonpolar,
       permitting the reactions of the organic compounds to occur within this environment.
       However, the enzyme as a whole must be soluble in water, the body's fluid. The polar side
       chains found on the outside of the enzyme sphere permit the enzyme to remain soluble in
       the body's fluids, while the nonpolar cavity provides an environment appropriate for the
       reaction of organic molecules.

APPENDIX A

Useful Mathematical Skills for Scientific Work

A.1 (a) 71.8   (b) 6.437   (c) 0.84   (d) 418
    (e) 0.27   (f) 13.8   (g) 450   (h) 1.0010
    (i) 7.6   (j) 14.5   (k) 7.66   (l) 10.000

A.2 (a) 0.375   (b) 0.312   (c) 0.647   (d) 0.600

A.3 (a) 3^5 (b) 6^7 (c) 7^3 (d) 10^1

A.4 (a) 4^8 (b) 7^4 (c) 3^{-3}

A.5 (a) 375   (b) 42,100   (c) 0.00739
    (d) 0.832   (e) 5,930,000   (f) 0.00000593

A.6 (a) 4.76×10^4 (b) 5.41×10^2 (c) 9.43×10^{-2}
    (d) 3.15×10^{-4} (e) 2.19×10^6 (f) 2.19×10^{-6}

A.7 (a) $(4 \times 10^6)(2 \times 10^{-4}) = 8 \times 10^2$

  (b) $\dfrac{2.2 \times 10^3}{5.5 \times 10^{-3}} = 0.40 \times 10^6 = 4.0 \times 10^5$

  (c) $\dfrac{(4.36 \times 10^2)(5.39 \times 10^{-3})}{(6.20 \times 10^{-2})} = 3.79 \times 10^1$

  (d) $\dfrac{(3.18 \times 10^{-4})(2.17 \times 10^1)}{(4.83 \times 10^3)} = 1.43 \times 10^{-6}$

A.8 (a) 5.48×10^{-4} (b) 7.95×10^{-5} (c) 2.45×10^{-2}

A.9 (a) $\dfrac{16 \text{ g}}{225 \text{ g}} \times 100\% = 7.1\%$ (b) $(250 \text{ g})(0.18) = 45 \text{ g}$

A.10 (a) $\dfrac{1.2}{44.0} \times 100\% = 2.7\%$ (b) $\dfrac{14}{602} \times 100\% = 2.3\%$

Answers to Photo Quiz

1. Both compounds that produce red contain strontium. A plausible hypothesis is; strontium will produce a red color in fireworks.

2. Chemical change; the substances in the test tube before and after have clearly different properties. A red substance (before) produced a silver substance and a gas (after).

3. There are six liquid layers. A seventh gaseous layer (air) is also present.

4. The number of moles and atoms in each of the beakers is the following:
   1.00 lb $C_{12}H_{22}O_{11}$ = 454 g $C_{12}H_{22}O_{11}$ = 1.33 mol $C_{12}H_{22}O_{11}$ = 3.60×10^{25} atoms
   1.00 lb NaCl = 454 g NaCl = 7.76 mol NaCl = 9.34×10^{24} atoms
   1.00 lb $NaHCO_3$ = 454 g $NaHCO_3$ = 5.40 mol $NaHCO_3$ = 1.95×10^{25} atoms
   Thus, 1 lb of table salt has the greatest number of moles, and 1 lb of table sugar has the greatest number of atoms.

5. There are approximately 4×10^7 atoms along a one centimeter length. Thus, an area of 1 cm^2 would contain approximately 1.6×10^{15} atoms.

6. The electrons in atoms exist only in certain discrete energy levels. When electrons fall from higher to lower energy levels, light is emitted with frequencies (colors) that correspond to the differences between the levels. Thus, only certain colors are given off, corresponding to the various energy differences that are possible. Each element has its own unique set of colors, as shown in the various spectra. The orange glow of a mercury lamp corresponds to the combination of colors given off as its electrons fall from higher to lower levels.

7. Electrons from the Van de Graf generator flow to all parts of the woman, including her hair. Since like charges repel, the strands of her hair are repelled from her body and from one another, causing them to stand on end.

8. As a general rule, the number of bonds and atom forms corresponds to its number of unpaired valence electrons: C, O, H, and Cl (draw dot structures) typically form 4, 2, 1, and 1 bond, respectively. Thus, the atoms in ball-and-stick model kits have holes drilled corresponding to those numbers. Since silicon is in the same chemical family as carbon, it, too, forms two bonds. Thus a carbon atom could be used to substitute for silicon in SiH_4.

9. Ammonium nitrate is NH_4NO_3, ammonium phosphate is $(NH_4)_3PO_4$, and potassium sulfate is K_2SO_4. Ammonium nitrate and ammonium phosphate would provide nitrogen. Ammonium phosphate would provide phosphorus. Potassium sulfate would provide sulfur. The name "iron sulfate" does not specify the oxidation state of iron, thereby leaving ambiguity as to the exact chemical formula of the compound present.

10. A 12.5 g sample of $(NH_4)_2Cr_2O_7$ (0.0496 mol) will produce 7.54 g of Cr_2O_3 (0.0496 mol).

11. Gases expand upon heating, thereby decreasing in density. Since less dense substances rise to the top of more dense substances, the hotter, less dense air inside the balloon rises above the cooler, more dense surrounding air, carrying the balloonist aloft. Air tends to be coolest at night and in the early morning, having its greatest density during these hours. These conditions provide the greatest density difference between the hot air in the balloon and the surroundings. Thus, a hot air balloon rises aloft most easily in the morning.

12. Whereas most substances contract upon cooling, water is unusual in its property of expanding as it freezes. This expansion results in a solid which is less dense than the liquid from which it freezes. Thus, ice floats in liquid water. Paraffin, which is typical of the more general behavior, contracts as it freezes, leading to a solid that is more dense than its liquid form. Thus, solid paraffin sinks in its liquid.

13. The solubilities of most solids are greater in solvents at high temperature than at low temperature. Sodium acetate in water exhibits this behavior. Thus, all of the solid dissolves in hot water. However, as the solution is allowed to cool, the solubility decreases. When the solubility of the substance drops below the concentration actually present in solution the excess solute is forced from the solution, forming an insoluble solid. In this case, slow cooling has permitted the solid to form beautiful needle-like crystals.

14. Acid rain is a severe problem in some parts of the world, including parts of the United States. Carbonates react with acids to form a salt, water, and carbon dioxide. In the case of insoluble carbonates such as marble ($CaCO_3$), the salt produced may be soluble, thereby dissolving in the acid. Thus, as acid rain washes over a marble statue, the statue dissolves slowly, as observed in the "before and "after" shots.

15. The more acidic a solution, the lower the pH. Thus, vinegar, with a pH of 2.5, is the most acidic.

16. The following oxidation-reduction is occurring: $Cu + 2\ Ag^+ \longrightarrow Cu^{2+} + 2\ Ag$. The metal formed from the reduction of silver ion is plating out on the copper coil. The blue color is from the copper(II) ion that forms as the copper metal is oxidized.

17. These photos illustrate Le Chatelier's principle. The reaction represented in the question describes the equilibrium between the blue $CoCl_4^{2-}$ ion and the pink $Co(H_2O)_6^{2+}$ ion. Addition of water drives the equilibrium to the right, converting blue $CoCl_4^{2-}$ ion to pink $Co(H_2O)_6^{2+}$. Addition of chloride ion drives the equilibrium to the left converting pink $Co(H_2O)_6^{2+}$ to blue $CoCl_4^{2-}$ ion.

18. Exposure to beta particles can cause damage to living tissue. Thus, beta emitters are not generally desirable for diagnostic purposes, where the goal is to assess the patient's condition. However, the destruction of living cells is often the goal of treatment, as is the case in reducing the size of an enlarged thyroid gland. Thus, a physician would select the milder iodine-123 for diagnosis, using iodine-131 only if the goal of treatment (as determined from the diagnosis) is to kill some of the cells present.

19. Organic compounds are carbon-containing covalent compounds, generally characterized by relatively low melting points. In contrast to this, inorganic substances are often ionic compounds with relatively high melting points. The substance on the right melts and then chars, leaving a carbon residue. Thus, it must be the organic compound. The substance on the left never melts under the same conditions. Thus, it must be the inorganic compound.

20. Celery is a polysaccharide, composed of repeating glucose units. Table sugar is a disaccharide, composed of a glucose and a fructose unit. Thus, both of these foods are carbohydrates. By contrast, nonfat milk is a protein, and butter is a lipid.